£9.95

SOCIETY FOR EXPERIMENTAL BIOLOGY

SEMINAR SERIES · 29

IMMUNOLOGY IN PLANT SCIENCE

IMMUNOLOGY IN PLANT SCIENCE

Edited by

TREVOR L. WANG

John Innes Institute, Norwich

CAMBRIDGE UNIVERSITY PRESS

Cambridge

London New York New Rochelle

Melbourne Sydney

Published by the Press Syndicate of the University of Cambridge
The Pitt Building, Trumpington Street, Cambridge CB2 1RP
32 East 57th Street, New York, NY 10022, USA
10 Stamford Road, Oakleigh, Melbourne 3166, Australia

First published 1986

Printed in Great Britain at the University Press, Cambridge

Library of Congress cataloging in publication data

Immunology in plant science.
(Seminar series/Society for Experimental Biology; 29)
Papers presented at the Leeds meeting of the Society for Experimental Biology in January 1985.
Includes index.

1. Plant immunology – Congresses. I. Wang, Trevor L.
II. Series: Seminar series (Society for Experimental Biology (Great Britain)); 29.
SB750.I455 1986 582.13′042 86–8215

British Library cataloguing in publication data

Immunology in plant science. – (Society for Experimental Biology seminar series; 29)

1. Plants – Disease and pest resistance
2. Micro-organisms, Pathogenic
I. Wang, Trevor L. II. Series
581.2′9 SB750

ISBN 307465 hard covers
ISBN 314801 paperback

CONTENTS

CONTRIBUTORS

Arnscheidt, A.
Fachbereich Biologie/Chemie, Universität Osnabrück, Postfach 4469, D-4500 Osnabrück, West Germany.

Atzorn, R.
Fachbereich Biologie/Chemie, Universität Osnabrück, Postfach 4469, D-4500 Osnabrück, West Germany.

Barbara, D. J.
East Malling Research Station, Maidstone, Kent, ME19 6BJ, U.K.

Brewin, N. J.
John Innes Institute, Colney Lane, Norwich, NR4 7UH, U.K.

Butcher, G. W.
Monoclonal Antibody Centre, AFRC Institute of Animal Physiology, Babraham, Cambridge, CB2 4AT, U.K.

Clayton, L.
John Innes Institute, Colney Lane, Norwich, NR4 7UH, U.K.

Clark, M.F.
East Malling Research Station, Maidstone, Kent, ME19 6BJ, U.K.

Cullimore, J. V.
Department of Biological Sciences, University of Warwick, Coventry, CV4 7AL, U.K.

Cuming, A. C.
Department of Genetics, University of Leeds, Leeds, LS2 9JT, U.K.

Doonan, J.
John Innes Institute, Colney Lane, Norwich, NR4 7UH, U.K.

Eberle, J.
Fachbereich Biologie/Chemie, Universität Osnabrück, Postfach 4469, D-4500 Osnabrück, West Germany.

Feyerabend, M.
Fachbereich Biologie/Chemie, Universität Osnabrück, Postfach 4469, D-4500 Osnabrück, West Germany.

Gadal, P.
Laboratoire de Physiologie Végétale Moléculaire, Université de Paris-Sud, Bâtiment 430, (U.A. CNRS 575), 91405 Orsay Cedex, France.

Galfrè, G.
Monoclonal Antibody Centre, AFRC Institute of Animal Physiology, Babraham, Cambridge, CB2 4AT, U.K.

Jordan, B. R.
Glasshouse Crops Research Institute, Rustington, Littlehampton, West Sussex, BN16 3PU, U.K.

Jourdan, P. S.
Fachbereich Biologie/Chemie, Universität Osnabrück, Postfach 4469, D-4500 Osnabrück, West Germany.

Larkins, A. P.
Monoclonal Antibody Centre, AFRC Institute of Animal Physiology, Babraham, Cambridge, CB2 4AT, U.K.

Mertens, R.
Fachbereich Biologie/Chemie, Universität Osnabrück, Postfach 4469, D-4500 Osnabrück, West Germany.

Partis, M. D.
Glasshouse Crops Research Institute, Rustington, Littlehampton, West Sussex, BN16 3PU, U.K.

Perrot-Rechenmann, C.
Laboratoire Structure et Métabolisme des Plantes, Université de Paris-Sud, Bâtiment 490, (U.A. CNRS 575) 91405 Orsay Cedex, France.

Roberts, K.
John Innes Institute, Colney Lane, Norwich, NR4 7UH, U.K.

Robertson, J. G.
John Innes Institute, Colney Lane, Norwich, NR4 7UH, U.K.

Thomas, B.
Glasshouse Crops Research Institute, Rustington, Littlehampton, West Sussex, BN16 3PU, U.K.

Weiler, E. W.
Fachbereich Biologie/Chemie, Universität Osnabrück, Postfach 4469, D-4500 Osnabrück, West Germany.

Wells, B.
John Innes Institute, Colney Lane, Norwich, NR4 7UH, U.K.

Wieczorek, U.
Fachbereich Biologie/Chemie, Universität Osnabrück, Postfach 4465, D-4500 Osnabrück, West Germany.

Williams, R. S.
Department of Biological Sciences, University of Warwick, Coventry, CV4 7AL, U.K.

Wood, E. A.
John Innes Institute, Colney Lane, Norwich, NR4 7UH, U.K.

PROTOCOLS

PREFACE

Very few texts pay attention to the many advances that have been made in the plant sciences through the application of antibody technology. This volume attempts to redress the balance and represents the invited papers presented at the Leeds meeting of the Society for Experimental Biology in January 1985. The meeting was designed to introduce the subject to those not familiar with immunology and to illustrate the range of subjects in plant biology which have benefited from immunological methods.

All bar one of the chapters are devoted entirely to plants and especially the practical aspects of using antibodies; each chapter being appended with self-contained step-by-step methods relevant to the topic in question. The initial chapter is the exception: it is directed at providing an overall introduction to immunology. It is the only chapter which does not contain a Protocols section *sensu strictu* since one could easily fill a book of this length with methods associated with one aspect of the chapter – the raising of monoclonal antibodies. Indeed, there are many volumes on precisely that subject. This chapter, however, does contain a useful scheme by which one can gauge the time and effort required to raise antibodies.

Several of the chapters cover general aspects of topics dealt with in detail elsewhere. Thus, the chapter on enzyme immunocytochemistry will be very useful for those interested in the chapters on the plant cell surface, the plant cytoskeleton and immunocytochemistry in general. Likewise, the details on the use of filter systems in the chapter on the plant cell surface complements the use of immunocytochemistry in the enzyme chapter and is a must for all users of fluorescence microscopy.

Finally, I must thank all the contributors for their excellent papers, Anne Williams for her devotion to duty against all opposition, including that from the word processor, and Angie for her tolerance and help with the reading of the manuscripts.

July 1985

T.L.W.
(S.E.B. Plant
Development Group)

ACKNOWLEDGEMENT

The editor would like to thank Shell Research Ltd, Sittingbourne Research Centre, Sittingbourne, Kent, on behalf of the Society for Experimental Biology, for its generous financial assistance with the organisation of the Leeds meeting at which these papers were presented.

G. GALFRÈ AND G. W. BUTCHER

Making antibodies

Introduction

People or animals who recover from certain kinds of infection are resistant to reinfection with the same pathogen. This "immunity" is highly specific – the resistance does not extend to pathogens other than the immunising one and may last for a very long time, i.e. decades. Only after the development of methods for the measurement of the state of immunity of an individual did immunisation become an important medical and veterinary practice. Today we are able to follow a particular immune reaction by directly monitoring either the response of the cells involved in that reaction, or the secreted products of these cells. Undoubtedly the simplest and most rewarding method has been to measure the concentration of circulating antibodies.

Although immunology evolved as a branch of medicine, it soon became clear that animals could be immunised against a wide variety of substances ("antigens") unrelated to pathogens. The methods for measuring antibody concentration have been adapted in various ways in order to measure antigen concentration. In this way the techniques of immunology acquired a practical importance for investigators in a wide variety of disciplines other than medicine. The special features of the immunological (serological) methods which make them so generally useful are the facility with which the antisera can be raised against widely different antigenic substances and the ability of the antibodies to pick out one molecular structure in a complex mixture.

Since the introduction of methods for the production of monoclonal antibodies (McAb) (Köhler & Milstein, 1975) the immunological approach has been extended in two important ways. First, specific antibody reagents can be derived without the need to achieve high purity of the antigen used for immunisation. Second, a given antibody can now be produced in large quantities at acceptable economic cost. In order to exploit the advantages of the immunological approach, a minimal knowledge of fundamental immunology would benefit even those researchers in fields remote from medical science.

Cellular aspects of the immune response

In fulfilling its role of defending the organism against infection the immune system must be able to recognise a wide variety of pathogens; at the same time, however, since the immune response is destructive, it must avoid recognising and reacting against itself. Since both the host and the potential invaders are composed of the same kinds of chemical structure – protein, carbohydrate, lipid, nucleic acids – the immune recognition must be finely tuned: a breakdown in the system will result in either fatal infection, or self-destruction.

The first requirement of an immune system which is to be capable of responding to a huge variety of antigens is a set of receptors with the capacity of discriminating precisely between all of them.

The evolution of a *small* family of molecular structures with these properties, although attractive in terms of genetic economy, may be impossible to achieve. The alternative solution, which has been adopted by the immune system of higher vertebrates, is the continuous random production of thousands of different receptor structures able to bind, more or less efficiently, any possible antigen. In the absence of the antigen each of these receptors is present only in small amounts. When antigen enters the system the receptors able to bind to it are selected for expansion. Each antigen-binding receptor is carried by a cell and each cell carries only one receptor. For example, in a human being, with as many as 2×10^{12} lymphocytes, a very wide spectrum of receptors can be represented at any one time. In this way the system is ready in advance to tackle any antigenic challenge. At the same time mechanisms are ready to suppress cells bearing receptors which bind to self molecules. This latter process of self-tolerance is not yet well understood and we will not go into a detailed account of it here.

The population of lymphocytes can be divided into two main lineages, the T cells, which depend for their normal development on the thymus gland, and the B cells, which do not. Both these cell types are descendants of pluripotent stem cells present in the bone marrow in adults. These two types of lymphocytes can be distinguished in practice by several differences in protein expression, particularly a number of cell-surface molecules. The most important characteristic of B cells is the expression of receptors on their surfaces classically identified as immunoglobulin (Ig), that can also be secreted in large quantities in soluble form by the terminally-differentiated B cell, the plasma cell. The T cell does not express immunoglobulin but carries some easily-identifiable specific marker proteins. Of course the most important, but historically very elusive, feature of the T cell surface is the presence of an antigen-specific receptor similar to but distinct from the Ig receptor of the B cell.

A very important physiological difference between T and B cells is the manner in which they are activated by antigen. Whereas B cells will respond under the appropriate circumstances to antigens which are in solution, T cells seem only to be activated by antigen presented on the surface of specialised antigen presenting cells. Whereas B cells seem to have but one function, i.e. the ultimate production of specific antibody, the T cell population contains cells which can perform a number of physiological roles e.g. the killing of infected or aberrant cells and the regulation of the immune response, either by the production of pharmacologically-active factors (lymphokines) or by direct cell-cell contact. In this chapter we will not be discussing the autonomous activities of T lymphocytes but rather will confine ourselves to a description of the processes involved in the production of antibody by B cells. Although some antigens can induce antibody production with little or no involvement of T cells (in particular, substances with a repetitive polysaccharide-based structure), for most antigens the presence of T cells is an absolute requirement.

The majority of immune responses are initiated in solid lymphoid tissue (*viz.* the lymph nodes and the spleen), where antigen is presented to immunocompetent lymphocytes. Foreign antigen reaches these sites via lymphatic vessels (in the case of lymph nodes) or via the bloodstream (in the case of the spleen). Both prior to and after reaching, say, a lymph node, antigen is taken up and processed by specialised lymphomyeloid cells (e.g. macrophages, "dendritic" cells). In a T lymphocyte-dependent response, a T "helper" cell bearing appropriate receptors is activated when it recognises antigen on the surface of one of these antigen-presenting cells. In the last decade a lot of effort has gone into the identification of the form in which the antigen is recognised by the T cell. Two features of this recognition of protein or glycoprotein antigens have been revealed. First, the T helper cell does not "see" antigen *per se* but sees antigen in some form of association with a "self" cell surface molecule, a class II histocompatibility molecule; second, an individual T cell is usually activated by a relatively short peptide derivative of the complete antigenic protein associated with this "self molecule" and not by the native form of the antigen. The binding of a T cell (in the G_0 phase of the cell cycle), carrying an appropriate receptor, to antigen in the form presented by a dendritic cell triggers differentiation and proliferation of the T cell to generate a clone of active helper cells. It is these cells which "help" antigen-specific B cells to make antibody. In parallel with T cell activation, B cells carrying surface receptors able to bind to the antigen in solution undergo the first steps in their differentiation sequence. This includes internalisation of receptor bound antigen followed by its processing. Although the intracellular events in antigen processing are probably very similar in B cells and dendritic cells, the uptake of antigen is specific in the case of B cells (being receptor-mediated), but non-specific for dendritic cells. In this way the same associations of antigen-derived peptides and class II molecules can be present on both the dendritic cells and the B cell. Specific helper T cells, which have previously been activated by binding to dendritic cells can "help" B cells when they recognise and bind to the same combination of peptide and class II molecule on the B lymphocyte surface. This cell-cell contact permits, by mechanisms unknown, the further differentiation of the B cell towards the terminal plasma cell. In consequence, a clone of T cells specific for one particular part of the antigen is able to help all the different B cells which can bind to any part of the antigen (Howard, 1985). This process is a satisfying explanation of the classical observation that one can obtain antibody response to small, intrinsically non-immunogenic molecules (haptens) by conjugating them to larger carrier proteins (Mitchison, 1971). B cells, which are able to bind to a free hapten cannot, however, proceed to terminal differentiation and antibody secretion, unless they receive the second, obligatory stimulus from a T cell. In the case of small hapten molecules, however, it seems that no appropriate form of the antigen can be presented, either by dendritic cells (to stimulate the resting T cell) or by a B cell (to receive "help" from the activated T cell). By coupling the hapten to a large protein, one is providing several T cell activating structures.

Receptor structures

Immunoglobulin

The antigen-binding receptor of a B lymphocyte is one of a family of proteins called immunoglobulins (Ig). These proteins can be inserted into the plasma membrane of the lymphocyte where they function as classical receptors to mediate the cellular response when binding to specific ligands, i.e. antigen. One result of this response is the differentiation of B cells into plasma cells which are highly specialised secretors of large quantities of the same receptor in a soluble form (serum Ig). Immunoglobulins are formed by the covalent linking of larger "heavy" polypeptide chains, (M.W. 50–65,000) with smaller "light" chains (M.W. ca. 25,000). The basic unit of Ig is a Y-shaped molecule consisting of two identical heavy-light (H-L) heterodimers in covalent association (M.W. 150–180,000) (Fig. 1). This unit is a bifunctional structure: antigen recognition is achieved by two identical antigen-binding sites situated at the ends of the arms of the Y. Both H and L chains contribute to the formation of the binding site. Interactions with other components of the immune system are mediated through sites on the tail portion of the molecule. This part of the molecule is formed by H chain sequences. In higher vertebrates the immunoglobulins can be categorised in five different classes based on their biochemical and biological properties (Table 1 and Fig. 2). These differences are due only to the structure of their heavy chains.

Table 1. *Properties of the immunoglobulins of higher vertebrates*

	Class of Ig				
	IgM	IgD	IgG	IgA	IgE
Molecular weight	900,000	185,000	150,000	160,000 or 335,000	200,000
Heavy chain	μ	δ	γ	α	ε
Light chain	κ or λ	κ or λ	κ or λ	κ or λ	κ or λ
Chain composition	$H_{10}L_{10}J$*	H_2L_2	H_2L_2	H_2L_2 or H_4L_4J	H_2L_2
% of total Ig in blood	5–10	<1	60–80	10–20	<5
Complement activation	+	–	+	–	–

*The J (joining) chain is a polypeptide of ca. 20,000 daltons which covalently links the heavy chains in IgM and IgA oligomers. It is totally unrelated to the J segments of Ig genes.

Fig. 1. *Schematic representation of an Ig molecule*
The tertiary structure of an Ig molecule is composed of a series of interacting Ig protein domains. The heavy chain domains $C_H{}^3$ and $C_H{}^2$ interact with like domains; $C_H{}^1$ and V_H interact with C_L and V_L respectively. Intrachain disulphide linkages (S – S) maintain domain structure, while interheavy and heavy-light disulphide bonds contribute to the stability of the Ig molecule.

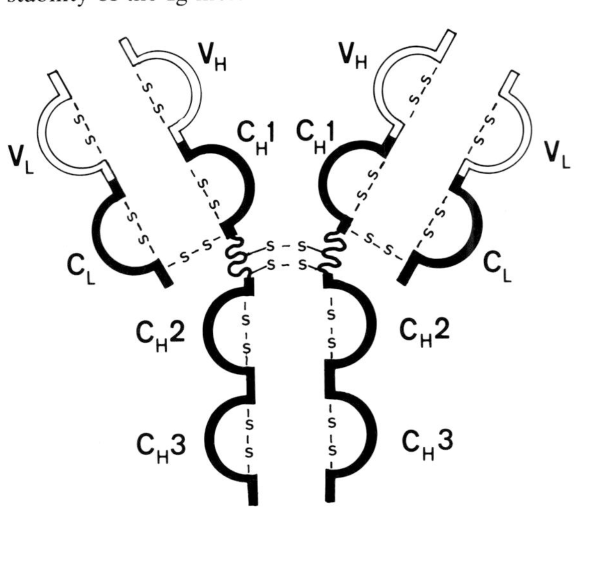

Fig. 2. *Three of the classes of Ig molecules*
The bifunctional IgG molecule is the simplest natural form of Ig. More complex structures are seen in dimeric IgA and pentameric IgM.

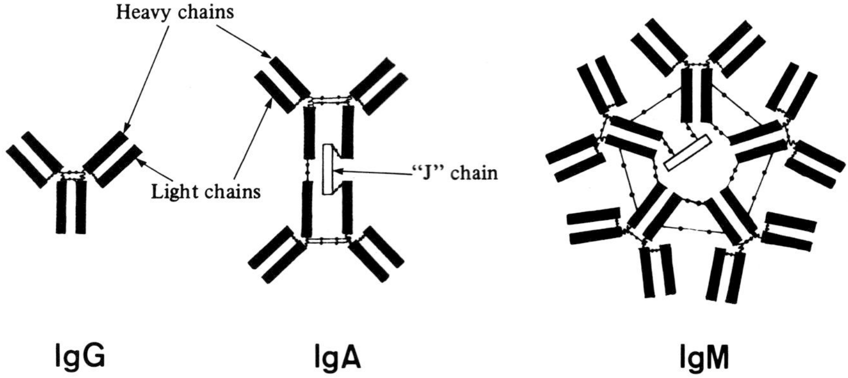

Light chains can also be classified into two types, and these are free to associate with any class of heavy chain. The most common immunoglobulins in serum are IgG. In most mammals IgG can be found in a number of forms called subclasses, which again are a consequence of differences in heavy chain structure. A notable exception to this is the rabbit which appears to possess only one form of IgG.

As we have just said, the bifunctionality of the Ig structure is a consequence of a basic asymmetry of all Ig polypeptide chains. In both heavy and light chains the N-terminal part will contribute to the formation of the antigen-binding site. Amino acid sequence analysis of a large number of immunoglobulins has shown that the N-terminal region is very variable: the differences between Ig chains of the same type are almost exclusively located within the first ca. 115 residues. This region is generally referred to as the variable or V region, while the rest of the polypeptide is known as the constant, or C region (see Fig. 1)

Immunoglobulin genes

The work of protein chemists in analysis of Ig chains made it clear that an individual is capable of generating a very large number of different B cell receptors. This number is so high that, if each molecule had to be coded for individually in the genome, the genetic load would be enormous. On the other hand, the alternative possibility of a small set of genes, highly unstable in somatic cells, has not been substantiated by studies of mutation rate in Ig-producing cell lines *in vitro*.

Only recently, with the application of gene cloning methods, the unravelling of the genetic complexity of the Ig genes has provided an adequate explanation for the generation of receptor diversity (Leder, 1982). In the mammalian genome there are three separate sets of genes for Ig chains, one coding for kappa light chains, another for lambda light chains and a third for all the types of heavy chains, each set being located on a separate chromosome. The genetic organisation of these three Ig gene regions is essentially similar, and we will therefore discuss only that for the heavy chains (Fig. 3).

Ig chains were the first example contradicting the old dogma "one gene: one polypeptide chain" in that each Ig chain is encoded by at least three

Fig. 3. *Organisation of immunoglobulin gene regions*
V and C refer to Variable and Constant gene segments; Js and Ds to clusters of "Joining" and "Diversity" gene segments. Note that the λ region, which contains very few V genes (2 in mouse), has a distinct arrangement but preserves the potential for V-J-C joining. The figure shows rearrangement for heavy chain genes leading to expression of IgM followed by a switch to IgG.

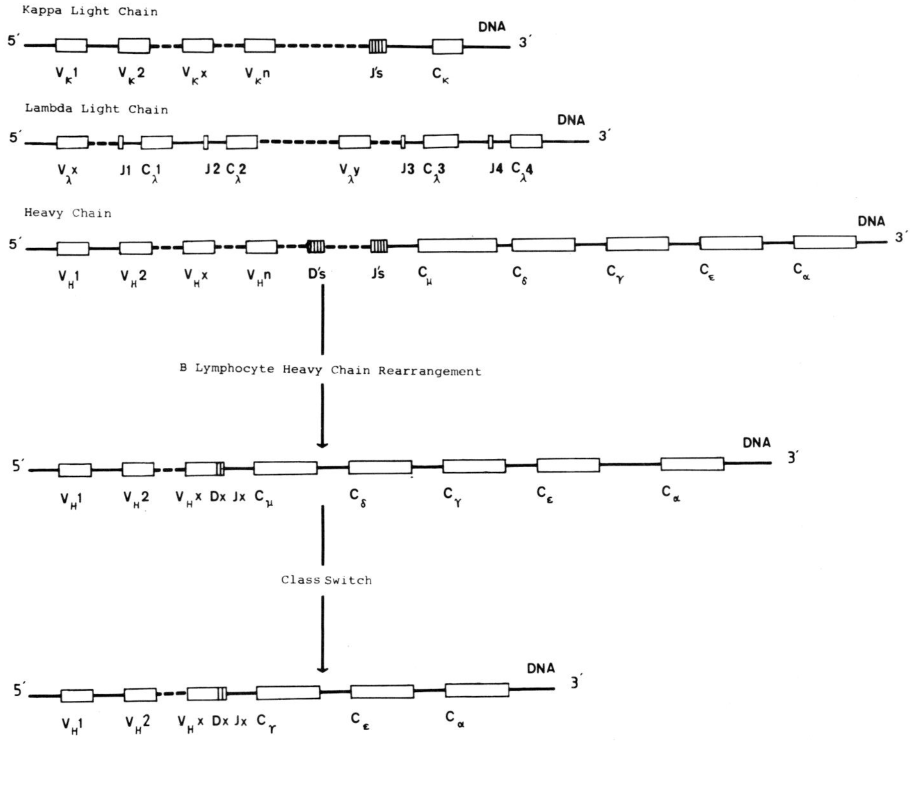

different genes. It is now clear that the configuration of Ig genes in B cells is different from that in any other cell type. In non-B cells there is a panel of several hundred different genes coding for part of the variable region (V genes) located several kilobases (kb) "upstream" of a smaller number (at least 10) "diversity" (D) genes. Several kb "downstream" of D is a set of a few (1–7) "joining" (J) genes which is in turn separated by a smaller distance from the regions coding for different heavy chain constant genes (Fig. 3). In the course of differentiation of every individual B cell, one V gene is joined to one D segment and one J segment and the intervening DNA is excised. Any of the V, D and J genes can be chosen for rearrangement, but once a successful VDJ combination has been produced, the cell seems to lose the ability for further V region rearrangement. In this way each B cell and the clone derived from it become committed to the expression of a unique binding site.

During B cell differentiation every successful VDJ combination is joined to the first (5′) constant region gene, Cμ. In later stages of differentiation of the cell the same VDJ combination can be moved next to the other heavy chain constant region genes e.g. Cδ, Cγ, Cε. Therefore within a single clone of B cells the commitment to the Ig binding site is greater than that to the effector portion of the molecule.

This complex genetic arrangement is a very economical way to create the large heterogeneity of receptors required by the immune system. There are four different contributions to the overall receptor diversity.

1. Germ line diversity: each of the V, D and J segments are represented by multiple, different copies in the genome.
2. Combinatorial diversity: the different V, D and J segments can be permutated for the production of V regions. Two different V regions (of the heavy and the light chain) combine to make a functional binding site.
3. Junctional diversity: the position of joining of the different gene segments can vary over a few nucleotides.
4. Somatic mutation: V regions are subject to somatic mutation, probably at a normal rate. Under the influence of selection by antigen this can be important in the maturation of receptor affinity during an immune response.

The T cell receptor

Recent advances have uncovered some of the features of the membrane-bound antigen-specific receptor of T cells. This receptor seems also to possess a two chain structure, and although the protein and DNA sequence information indicates that both chains are non-immunoglobulin, they are nevertheless clearly of the 'Ig family' of proteins, which includes other cell-surface molecules such as Thy-1 and the class I and class II histocompatibility antigens (Williams, 1984). Genetic and molecular biological studies further indicate that the T cell receptor genes, while being distinct from the Ig genes, also undergo the kinds of somatic rearrangement which we have described above for B cells and so probably employ similar strategies for the generation of receptor diversity. It is, however, too early to predict how close the parallel between B and T cell differentiation will turn out to be.

Maturation of the immune response

During clonal expansion of an antigen-specific B cell a series of morphological events take place to transform the resting lymphocyte, which has little cytoplasmic activity, into the antibody-secreting plasma cell, with a high ratio of cytoplasmic to nuclear material. These events are necessary to build up the protein-synthetic machinery of the cell, e.g. development of the rough endoplasmic reticulum and Golgi apparatus. More subtly, there are frequently also changes in the immunoglobulin expressed by the cells as a result of changes in the genomic regions coding for the immunoglobulins. The phenomenon concerned is known as "class-switching", because it involves the movement of the previously rearranged VDJ combination from its position near the Cμ gene to an equivalent position upstream of one of the other heavy chain region genes (Fig. 3). Thus, in a common case, a B cell producing antibody of the IgM class against a given antigen will generate progeny producing antibody of the IgG class with the identical combining site.

A property of the immune response which we have not so far considered is that of immunological "memory". When an animal is challenged with an antigen against which it has previously reacted, its response is generally more rapid and more potent than on the first occasion. This phenomenon is specific: the animal can somehow remember the first antigenic challenge and is primed for subsequent contact with the same antigen. The initial encounter with antigen will usually lead to the expansion of specific T and B cell clones. Simple clonal expansion, however, is probably not adequate to explain the longevity and specificity of immune memory: instead the generation of a separate class of specialised "memory" lymphocytes seems most likely. Although the mechanism of memory cell generation is not fully understood it is presumed that the cells arise as side-lineage(s) during antigen-dependent B and T cell clonal expansion.

In the course of specific immune response, during clonal expansion and differentiation in both primary and secondary responses, there is a continuous maturation of the antibody response towards high affinity receptors. Since B cell activation is antigen-dependent, affinity maturation is likely to result from competition between different B cell clones for limiting quantities of antigen. Occasional somatic variants, if they express receptors of higher affinity than their progenitors, can also be selected by this same process. The production of high affinity antisera by long-term immunisation schedules takes advantage of this process.

Derivation of hybrids secreting specific antibodies

Myelomas

By physicochemical methods it is possible to separate the Ig fraction of a serum to its component classes and, with some difficulties, subclasses, but the separation of individual Ig, differing only for the variable region, is generally impossible. Tumours of Ig producing cells occur spontaneously in many mammals. They quite frequently arise in man and in particular strains of mice and rats. The associated pathological disorder was originally classified as "multiple myeloma" in man, because multiple tumours develop in the bone marrow (or myelogenous tissue). The disease is generally diagnosed by the appearance in serum and urine of particular proteins, called "myeloma proteins" easily detectable by electrophoresis. Although these proteins were described by Bence Jones in the nineteenth century, only in the 1950's was their relationship with antibodies recognised, and the cell type responsible for their secretion identified as the plasma cell (hence the introduction of the name of plasmacytoma). An essential characteristic of multiple myelomas is the appearance of tumours in several sites of the affected animal, but it has been demonstrated that all of these actually derive from the uncontrolled expansion of one clone, or occasionally a few clones, of plasma cells. The tumour, therefore, is generally monoclonal in origin and the term monoclonal is now used when referring to the myeloma protein as well as to the cells. Furthermore, monoclonality can be ensured in any case in which the malignant cells can be grown *in vitro* and where they can be subcloned. Nearly all we know about Ig structure and genetics comes from the study of either naturally occurring or chemically induced myelomas (Porter, 1973; Kabat, 1976).

Over the years a number of immunologists have made considerable efforts to find myeloma proteins with a defined specificity (i.e. antibody activity). Attempts were made to induce specific tumours with carcinogenic chemicals in immunised animals. Alternatively, already existing myelomas were tested for their ability to bind to a wide variety of substances (natural or synthetic) known to be good antigens. Another method, the induction of B cell tumours by certain viruses, was more successful but the tumour lines thus derived are often difficult to handle, because of their poor growth characteristics *in vitro*. Despite great efforts progress along these lines was very slow. Indeed, the introduction of the highly efficient and robust cell fusion method for specific antibody derivation swept the other approaches aside in the realm of experimental immunology. We should mention, however, that in the case of the production of human McAb, the use of *in vitro* Epstein-Barr virus transformation of B lymphocytes is currently, and may remain, a popular technique.

Somatic cell fusion

Hybrids produced by somatic cell fusion initially possess two sets of chromosomes, one set derived from each of the parent cells. During cell division some of the chromosomes may be unable to duplicate properly and are lost. After a number of cell cycles, however, a more stable reduced karyotype is achieved. In interspecific hybrids this reduction process is not random but involves preferential loss of one of the parental species' chromosomes. Best documented combinations are human × mouse, human × Chinese hamster and mouse × rat hybrids from which, respectively, the human, the Chinese hamster and the rat chromosomes are preferentially lost.

The phenomenon of differential chromosomal loss has been extensively exploited for mapping of phenotypic characters to different chromosomes and to study gene regulation and dominance. The potential of somatic cell fusion to immortalise in the permanently growing cell line a phenotype otherwise only transiently expressed in normal cells, was not fully realised before the experiment of Köhler and Milstein in 1975. When a tumour cell is fused with a normal cell, hybrids can be derived with growth characteristics of the tumour and expressing some of the phenotypic characters of the differentiated normal cell.

The production of a protein by a cell depends not only upon the presence of the structural gene for that protein but also upon appropriate regulation. Thus, in a hybrid cell, the expression of a phenotype derived from one parent may or may not be apparent depending upon the nature or state of the other parent. For example, hybrids derived from myelomas and fibroblasts never express Ig while, on the contrary, hybrids between myelomas of two different species like mouse and rat secrete Igs of both parents (codominant expression) (Cotton & Milstein, 1973). The availability of cell lines suitable for fusion has played an essential role in the development of the methodology for the derivation of hybrids secreting McAb.

HAT selection

When using a permanently dividing cell line as a parent for hybrid production one needs to select the newly formed hybrids from the parent. The usual way to achieve this is to use parents carrying markers, such as an enzyme deficiency or a drug resistance. The most commonly used selection system is called "HAT selection" from the use of culture media supplemented with hypoxanthine, aminopterin and thymidine. The method was first described by Szybalska & Szybalsky (1962) and later by Littlefield (1964). Originally it was used as a classic complementation system, in which one cell lacks the enzyme hypoxanthine-guanine phosphoribosyl transferase (HGPRT, E.C.2.4.2.8) and is therefore unable to make use of hypoxanthine for synthesis of purines, and the other cell lacks a different enzyme (thymidine kinase, TK, E.C.2.7.1.21) and cannot utilise thymidine. Both types of cell are killed by aminopterin, a potent inhibitor of *de novo* synthesis of DNA precursors. Hybrids, on the other hand, can easily survive by complementation of the enzymatic activities from the two parents that enable them to use both hypoxanthine and thymidine.

For McAb derivation the HAT system is used in a similar way, but here one of the parents, the myeloma, is lacking either HGPRT or TK and is unable to grow in HAT, while the other parent, the lymphocyte, is unable to grow *in vitro* for more than a few cell cycles. Hybrids are able to grow in these conditions because they inherit from the myeloma parent the ability to divide *in vitro* and from the lymphocyte the correct enzymatic complement to grow in HAT.

Selectivity of somatic cell fusion

To derive hybrid myelomas a large number of spleen cells from an immunised animal are fused with cells of a suitable myeloma line by the use of polyethylene glycol (PEG). The population of cells obtained by mechanical disruption of a spleen contains a mixture of different cell types. Of the lymphocytes about 40% are T cells of different sorts and the remaining 60% include all the differentiation stage of B cells from pre-B cells to mature plasma cells and memory cells. At any one time less than one cell in a thousand in an immunised spleen produces Ab to the antigen used for the immunisation, and less than 5% secretes Ig of any sort. In contrast the population of hybrids derived by fusing spleen cells with a myeloma cell line is much more homogeneous in that none of the hybrids appear to have T cell characteristics, and 60–90% of them secrete Ig. If, instead of a myeloma, a T cell lymphoma is used for fusions with spleen cells the resulting hybrids bear T cell markers and none secrete Ig (Köhler, Pearson & Milstein, 1977). It appears therefore, that the phenotype of the tumour parent is a major selective influence on the outcome of cell fusion experiments.

Planning of the fusion experiment

The goal of a cell fusion experiment is the immortalisation of a cell secreting a specific antibody. The overall design of the experiment is schematically represented in Fig. 4. At each step of the procedure (immunisation, fusion, screening etc.) multiple choice of methods and approaches are possible and careful planning is important for success. The immunisation of an animal with an antigen increases the number of cells in the spleen expressing a specific antibody. The state of immunisation of the animal is therefore an important factor in determining the success of the fusion and so in any experiment it is sensible to test the serum of the animals for the presence of the specific antibody and to choose those animals making the strongest response.

Fig. 4. *Schematic protocol for the derivation of hybrid myelomas*

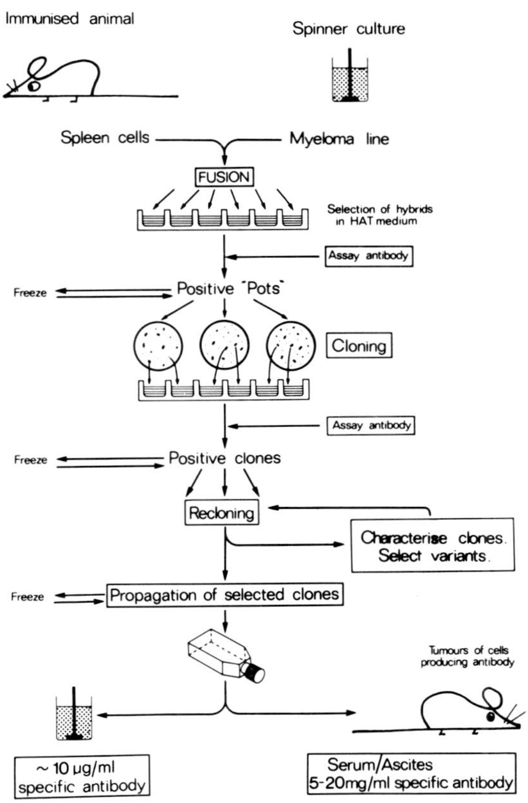

Unless an effective immunisation protocol is already known for the antigen of interest, it is advisable to inject a number of animals, and preferably animals of different species (mice, rats) or strains. The use of inbred strains of animal is advisable to avoid problems in the production of ascites where histocompatibility between tumour and host is necessary. The more significant scientific reason for choosing to work with rats or mice is the relative responsiveness of these animals to the antigen. In many cases both species respond well and the choice is made on other grounds. The choice of species is more or less restricted to mouse and rat for which *in vitro* growing plasmacytomas of high fusion potential are available. Cross species fusions can be done, and are reasonably successful between mouse and rat and, between both Syrian and Armenian hamsters and mouse (Sanchez-Madrid, Szklut & Springer, 1983). But with increasing phylogenetic distance the problems of karyotypic instability and low fusion efficiency seems to increase e.g. with rabbit (Yarmush *et al.*, 1980) and with man. Where practical, however, intraspecies fusions are to be performed. A list of mouse and rat plasmacytomas in general use is given in Table 2. As mentioned there, some of these tumour cell lines still produce their own Ig chains and, therefore, the hybrids made from them secrete not only the specific, spleen-cell derived antibody but also molecules

Table 2. *Myeloma cell lines used for hybridoma derivation*

Name	Strain	Immunoglobulin expression	Expected expression in hybrids	References
Mouse lines				
P3-X63Ag.8	BALB/c	MOPC 21 IgG_1 (κ)	HLGK	a
X63Ag8.653	BALB/c	None	HL	b
NSI/1Ag4.1	BALB/c	MOPC 21 κ (non-secreted)	HLK	c
NSO/U	BALB/c	None	HL	d
Sp2/0-AG14	BALB/c	None	HL	e
MPC11.45.6Tg1.7	BALB/c	MPC11 IgG_{2b} (κ)	HLGK	f
MPC11.45.6Tg1.7.5	BALB/c	MPC11 κ	HLK	g
S194/5.XX0.Bul	BALB/c	None	HL	h
Rat lines				
IR983F	LOU	None	HL	i
Y3Ag1.2.3	LOU	S210 κ chain	HLK	j
YB2/0	(LOU × AO)F_1	None	HL	k

For explanation of H, L, G and K terminology see Figure 5 and the text. References: a – Köhler & Milstein (1975); b – Kearney *et al.* (1979); c – Köhler, Howe & Milstein (1976); d – Galfré & Milstein (1981); e – Shulman, Wilde & Köhler (1978); f – Gefter, Margulies & Scharff (1977); g – Sharon (1980); h – Trowbridge (1978); i – Bazin (1982); j – Galfré, Milstein & Wright (1979); k – Kilmartin, Wright & Milstein (1982).

built up by association of Ig chains from both of the hybrids' parents. From cases such as this hybrid molecules are being produced, it is possible to derive variants which no longer express one of the chains. The pattern of Ig chain loss is not random but usually follows the rule that the number of different heavy chains produced is equal to or less than the number of different light chains (see Fig. 5). For general use the non-secreting plasmacytomas are preferable.

Finally, in this consideration of the planning of the fusion experiment, it is worth stressing an issue of practical importance. Before entering into the experiment itself one is at an advantage if one has devised an assay system which can test for antibody activity in hundreds of hybrid culture supernatants per day. The assay, has, therefore, to be fast, reliable and applicable to small samples. A variety of immunological assays are available for the detection of antigen-antibody reactions and each type of assay will detect some types of antibody better than others (see Protocols index and later chapters). For example, a conventional radioimmunoassay using a labelled ligand will work better with high affinity antibodies, while an agglutination

Fig. 5. *Chain composition of hybrid myelomas*
Hybrid myeloma cells synthesise not only the heavy (H) and light (L) immunoglobulin chains derived from the spleen cell, but also the heavy (G) and light (K) chains of myeloma origin. From hybrids expressing the four Ig chains (HLGK) variants lacking the expression of some of them can be isolated. Derivation of hybrids secreting only the specific antibody (HL) can be more easily prepared using mutant myelomas that have already lost one or both immunoglobulin chains. (+) stands for secretion of specific antibodies. Note that HL hybrids also spontaneously lose expression of Ig chains and therefore of Ab activity, hence the need to periodically reclone McAb producing hybrids.

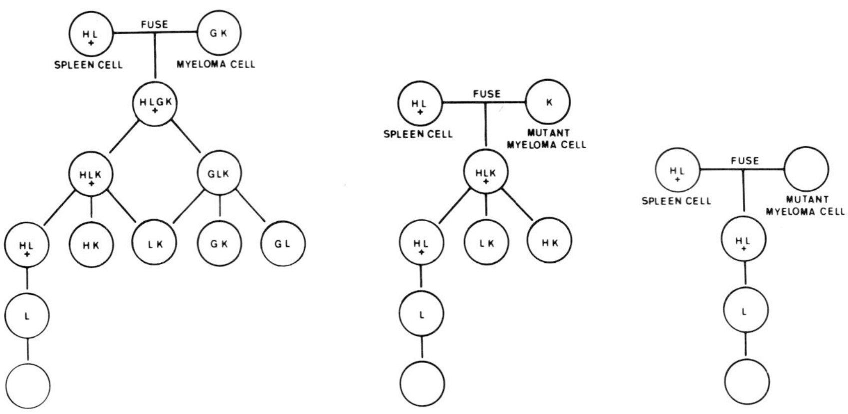

test will depend more on multivalent binding and so will detect preferentially IgM antibodies. It is advisable, therefore, to design the screening system for the fusion to reflect the anticipated use of the specific antibody, if this is compatible with assay speed.

Diary of a successful experiment

In this section we describe the derivation of a set of McAb as it was actually done in our laboratory. This project was the result of a collaboration between the Monoclonal Antibody Centre and a group of scientists at Long Ashton Research Station (LARS); the scientific results are published elsewhere (Notton, Fido & Galfrè, 1985). Here we intend to illustrate the time scale involved in such a project and to give some examples of the difficulties encountered and some possible approaches to their solution. A wide variety of methods were used and they have all been described elsewhere (Galfrè & Milstein, 1981) – here we will only discuss the principles involved in the method used in order to justify our choices.

1. Antigen preparation

The Ag in question, higher plant nitrate reductase, is a complex molecule formed of two subunits of ca. M.W. 100,000 and a molybdenum containing co-factor in the form of the pterin complex. Purification of the entire complex has been possible in only a few species and one of the uses of McAb will be the improvement of purification methods. The LARS team had developed good methods of purification before the project started. In general for McAb derivation purification of the Ag is not an essential prerequisite. In fact probably the most important aspect of the McAb technology is the ability to derive Ab to Ag which had been impossible to purify (a classic example being interferon; Secher & Burke, 1980) or that have not even been described before, as a very large variety of literature about cell surface antigens illustrates (Williams, Galfrè & Milstein, 1977). Nevertheless, wherever possible, purification of the Ag at least to some extent will greatly help. It is worth pointing out that chemical purity and immunological purity do not necessarily coincide. An Ag chemically contaminated with substances that are not antigenic in a particular animal, is in fact immunologically pure in that particular context. The purity of the Ag is only important in two considerations:

(a) If the impurity is more antigenic than the Ag itself, or able in any way to reduce the specific immune response.
(b) The method used to test for the response must be capable of distinguishing between the presence of Ab to the Ag and Ab to the impurity.

2. Planning the immunisation and experiment

Day minus 83: Four female F344 rats were injected subcutaneously with 200 μl (200 μg) of Ag dissolved in buffer and emulsified in complete Freund's adjuvant (CFA). The choice of animals has already been discussed. Here we did not even try mice as we expected an immediate good response; the Ag is a large plant protein and therefore is likely to be a strong immunogen. Apart from their availability at that time in our animal house, F344 rats were chosen because this inbred strain carries a particular allotype of the Igk light chain, called "b", that is different from that carried by the myeloma Y3 that we intended to use. This is a mutant myeloma that only secretes light chain (K) and therefore Ab secreting hybrids derived from it, are expected to express not only the specific heavy (H) and light (L) immunoglobulin chains of spleenocyte origin, but also the light chain (K) of the myeloma (see Fig. 5). The allotypic difference in light chains between F344 rats and LOU rats (from which Y3 was originally derived) offers a tool to differentiate the two light chains secreted by the hybrids.

Day minus 56: the animals were boosted by a subcutaneous injection of 200 μl (200 μg) of Ag dissolved in buffer and emulsified in incomplete Freund's adjuvant (IFA).

Day minus 44: each animal was tail bled and the serum was tested for the presence of specific Ab. This first bleed is used not only to monitor the immune state of the animal, but also, and perhaps more importantly, to set up and optimise the assay for specific Ab. In this example we decided to adopt a double screening procedure: all our antisera and hybrid supernatants would be assayed for their ability (a) to bind to purified Ag in an ELISA assay and (b) to inhibit specific enzymatic activity of the antigen. The ELISA assay consists of coating plastic microtitre plates with the purified Ag. Antisera or hybrid supernatants are then incubated in each well, the excess unbound Ab washed away, and the presence of bound Ab is revealed by the binding of a second, enzyme-labelled antiserum specific for rat Ig. Properly set up, this assay is extremely fast and reliable – 400–500 samples can be easily tested in 8 hours. The problem is that it requires Ag pure enough to coat the plate. The process of binding to plastic is a more or less non-specific step: most proteins will stick to the plastic, generally in direct proportion to their concentration in the coating solution. If the same impure preparation of Ag is used for immunisation and for coating of the ELISA plate, the assay will not distinguish between Ab to the Ag and Ab to the impurity if these are present at a sufficiently high level. In our case the Ag was 90% chemically pure and therefore we expected very few false positives.

The second assay, inhibition of enzymatic activity, is strictly related to the biological properties of our Ag, and is a particularly interesting assay because, although more difficult and time consuming than the ELISA (in 8 hours only a few dozen assays could be performed) it immediately gave information about the possible binding sites of the Ab along the Ag molecule and the reassuring feeling, when positive, that we were getting a response against the right specificities. In addition, it could be performed with crude plant extracts containing the Ag. The sera from the first bleeds of our animals were all positive in the enzyme inhibition assay, the worst at a dilution of about 1 in 200 and the best at 1 in 1200.

Day minus 31: second boost of the animals exactly as on Day minus 56.

Day minus 21: second bleed and test of sera. All the sera appear to be positive, but they were all now positive only to a dilution of 1 in 500. We never found a good explanation for this and at the time our ELISA assay was not working reliably because of a problem in antigen purification. Nevertheless, we decided to proceed with rat no. 1 as the immunisation seemed acceptable. It is worth noting that in many cases one does not really have experience of what to expect in terms of antiserum titres in rats or mice since conventional antisera are usually prepared in rabbits.

Day minus 5: to increase fusion efficiency it is essential that the myeloma cells are kept in exponential growth for a few days before fusion. Myeloma cultures are generally fed every other day in our laboratory and not during weekends, but a few days before fusion special care should be taken to get them in top shape. If myeloma cell lines are not kept growing continuously in the laboratory, then cultures should be started at least 10 days before fusion.

Day minus 3: The selected animal, and a second reserve one, are boosted with an intravenous injection of 100 μl (100 μg) of Ag dissolved in physiological saline.

3. Cell fusion and selection of hybrids

Day 0: the animal selected and injected on Day minus 3 is killed, the spleen is removed and cells are used for fusion according to our standard protocol. Half the spleen cells are fused with the rat myeloma Y3Ag1.2.3 (or Y3 for short) and half with the myeloma YB2/0 (or YO for short). This is generally done because fusion experiments are not very reproducible and it is wise to work with different cell lines to avoid heavy losses in case of contamination. After PEG treatment the cells are seeded into 4, 24-well plates (2 cm^3 $well^{-1}$) in dilutions such that about 4×10^6 spleen cells $well^{-1}$ are seeded into the wells of the first plate, 2×10^6 $well^{-1}$ in the second, 10^6 $well^{-1}$ in the third and 5×10^5 $well^{-1}$ in the last. Each well is numbered individually so that we have wells from 1/1 to 1/96 and from 2/1 to 2/96.

Day 1: all the plates are carefully checked for contamination and number of cells growing. Since we feed with normal medium on Day 0 the parent myeloma cells are dividing normally. Half the supernatant is removed from each well and replaced with selective medium containing HAT.

Day 2: parental myeloma cells show signs of dying. Plates are fed as at Day 1 with HAT medium.

Day 4: feed with HAT.

Day 7: all parental myelomas dead. Few if any, cells alive. Feed with HAT.

Day 11: foci of cells growing in some of the wells in high concentration plates (no. 1) may be hybrids. Feed all with HAT.

Day 14: hybrids obviously growing in at least 25% of the wells. Feed all with HAT.

Day 18: several wells (33 out of 192) now have confluent cell growth. One of the plates had several wells contaminated with fungus and, as none of the remaining wells showed signs of growing hybrids (this was a plate seeded at high dilution) the plate was discarded to avoid spreading the infection. Supernatants are harvested for testing and half the cells are removed and transferred to "duplicate plates". Original plates are fed with HAT medium and the duplicates with medium containing only HT. From now on plates are fed twice weekly and if new wells become confluent in the original plates, duplicates are prepared.Duplicate wells keep the same number as the original well because they presumably contain the same cell population (polyclonal at this stage) and we identify original wells as "cell lines". Different wells or tubes or flasks that are prepared from the same "line" are given the line name, e.g. 1/26, followed by some other specification in brackets, like plate number or tube number, e.g. 1/26 (P2); 1/26 (P6) and 1/26 (T1).

4. Screening and selection of McAb secreting lines

Day 19–20: of the supernatants tested more than half are found to be positive by ELISA and of these only half (9/33) show any ability to inhibit the enzyme activity of the Ag (nitrate reductase). At this stage we have many more hybrids from the Y3 fusion (25) than from the Y0 fusion (8) but the distribution of positives was normal (17/25 and 3/8). Later more hybrids appear in both fusions and we end up with more or less the same number of interesting lines from each fusion.

Day 21: all confluent wells are fed by removing half of the cells and adding fresh medium. The cells removed are generally discarded unless duplicates need to be prepared, either because the line is an "interesting" one (and therefore deserving of special treatment) or because duplicates do not exist (not yet made or lost because of contamination).

Day 25: supernatants from all confluent wells are harvested for assay. A dozen wells are lost because of bacterial contamination, but duplicates already existed. New duplicates are prepared where necessary. All plates are fed.

Day 26: ELISA and enzyme inhibition assay. More than half the "lines" tested are positive by ELISA, and about a quarter are also positive for enzyme inhibition.

Day 27: five lines are chosen as the most interesting positives comparing the results from the two screens already done (at Day 20 and Day 26) and cells from actively growing wells already scored twice as positive are cloned in soft agar. For a dozen of the positive lines separate cultures are prepared by transferring cells from confluent wells into flat sided 10 cm^3 tissue culture tubes.

Day 28: all plates are fed as usual.

Day 29: several of the tubes prepared on Day 27 are confluent, supernatants are separated by centrifugation and the cells are frozen in vials for liquid nitrogen storage. From now on, frozen samples of positive cultures are prepared by this method whenever cells are available in sufficient numbers.

Day 32: all plates are fed. Supernatant from all confluent cultures are collected for testing. Wells without actively growing cells are drained and washed with alcohol (about half of the original wells). Of the five lines cloned on Day 27 four are contaminated by bacteria and the agar plates have to be discarded.

Day 33: on the basis of new tests a second attempt at agar cloning is made for six lines (three already attempted at Day 27 and three new ones).

Day 35: all plates are fed.

Day 39: all plates are fed and supernatants collected for testing.

Day 41: for each of the seven lines cloned (1 Day 27 and 6 Day 33) 48 individual clones are picked from the agar plate using Pasteur pipettes. Each clone is transferred into a 2 cm^3 well of a 24-well plate and fed with medium containing 20% FCS but neither HAT nor HT. A total of 14 new plates are therefore prepared. Each pair of plates containing clones of the same line is labelled with the line name and each well is numbered (from 1 to 48). Individual clones will be identified, therefore, by their "line name" and clone number e.g. from line 1/26 prepare clones 1/26.1 to 1/26.48.

Day 42: all the original and duplicate plates are fed (not the ones containing clones).

Day 46: all plates are fed (not the ones containing clones).

Day 49: as Day 46.

Day 53: several of the clones picked on Day 41 have become confluent. Supernatants from all confluent cultures are collected for screening and all plates are fed. A dozen wells in the original plates are lost from yeast contamination.

Day 56: some original culture of uncloned lines that are still positive are transferred from wells to 25 cm^2 tissue culture flasks for further expansion; all the original and duplicate plates are then stopped. For each of the clone lines, three positive clones are chosen to be kept; for these, duplicate cultures are prepared by transferring half of the cells from the wells to 25 cm^2 tissue culture flasks. All plates containing clones are fed.

Day 57: the flasks prepared on Day 56 are examined for contamination and cell growth. If necessary, more fresh medium containing 10% FCS is added (to a maximum of 20 cm^3).

Day 60: all the plates are fed, supernatants are collected for testing only from wells containing clones chosen as positive. Flasks are examined for cell growth and, in the case of clones, whenever possible duplicate flasks are prepared. For each cloned line (six of them) we have now three selected clones and for each clone we have a growing 2 cm^3 well and two 25 cm^2 flasks. We keep feeding the wells with medium supplemented with 20% FCS, but the flasks are fed first with 10% FCS and later, if the cells grow well, with 5% and then 2.5% FCS. To keep duplicates and feed them with a medium of different composition and therefore from different batches of bottles, is the only way to avoid losses due to contamination.

Day 61: tests of supernatants show that we have succeeded in deriving six different McAb.

Supernatants from sister clones can now be pooled and more intensive characterisation work gets under way. Of course, some preliminary experiments on the specificity of the Ab have already been done, but up to now it has not been wise to invest too much effort in any of the lines as they have been neither monoclonal nor certain of survival.

Conclusion of the experiment

From Day 62 onwards, the experiment proceeded as follows. As soon as was possible we prepared at least six frozen samples for each clone. Cells from each clone were grown for 18 hours in a medium containing a radiolabelled amino acid and spent supernatants were analysed by SDS polyacrylamide electrophoresis. Two out of three clones for each line were then discontinued and only one "master" clone per line was kept growing. Only later, when more information was available about the specificity of the different McAb, cells from some of the cloned lines were injected into the appropriate rats for tumour generation and ascites production. When sufficient stocks of supernatant and ascites had been prepared, growth of all the lines was discontinued, frozen cell stocks being available whenever (or if ever) stocks of Ab run out.

A few more hybridomas different from the first six whose derivation has been described here could probably be prepared if necessary by thawing all the samples frozen between Day 29 and Day 60 and cloning them, but before attempting more cloning we like to study the supernatants stored to identify different and useful McAb. Thawing and cloning lines frozen at early stages of the experiment is not always successful as only a small percentage of cells are actually recovered from liquid nitrogen storage and, therefore, more intensive cloning is sometimes required.

Thus nine weeks after the cell fusion had been performed, a set of six

McAb selected for their differential abilities to inhibit the various enzymatic activities of nitrate reductase, were cloned and reasonably secure. This, of course, had been a particularly smooth-running experiment in which none of the common snags of this kind of work (e.g. contamination of cultures and assay breakdowns) had interfered.

The final events in the story are as follows. First, once it is clear that these cell lines were to be included in a published paper, they were given "Institutional" names, in this case being renamed AFRC MAC 75 to 79. Second, as is now our routine procedure, the cell lines were given an extra layer of security by being deposited in the National Collection of Animal Cell Cultures at Porton Down, Wiltshire, U.K. This is an important safeguard against accidental breakdown of our liquid nitrogen storage system.

The last 10 years has seen the application of McAb to all branches of animal physiology and biochemistry and has seen enormous benefits from the new approach. It is now apparent that the potential value of McAb to plant science is being recognised – a fact exemplified by the contributions to this volume. Although there may be particular problems to be faced, such as the extreme and, therefore, dominating immunogenicity of certain plant glycosidic structures, we can look forward to a flourishing relationship between those unexpected partners, immunology and botany.

KEY REFERENCES

Alberts, B., Bray, D., Lewis, J., Raff, M., Roberts, K. & Watson, J. H. (1983). *Molecular Biology of the Cell*, chapter 17, pp. 952–1012. New York: Garland Publishing.

Hood, L. E., Weissman, I. L., Wood, W. B. & Wilson, J. H. (1984). *Immunology*, 2nd edition. Menlo Park, California: Benjamin/Cummings Publishing.

Hudson, L. & Hay, F. C. (1980). *Practical Immunology*, 2nd edition. Oxford: Blackwell.

Köhler, G. & Milstein, C. (1975). Continuous cultures of fused cells secreting antibody of predefined specificity. *Nature*, **256**, 595–7.

Also several authors in *Methods in Enzymology*, vol. 73, part B, eds. J. J. Langone & H. Van Vunakis. New York: Academic Press.

REFERENCES

Bazin, H. (1982). Production of rat monoclonal antibodies with the LOU rat non-secreting IR983F myeloma cell line. In: *Protides of the Biological Fluids 29th Colloquium 1981*, ed. H. Peeters. Oxford and New York: Pergamon Press.

Cotton, R. G. H. & Milstein, C. (1973). Fusion of two immunoglobulin-producing myeloma cells. *Nature*, **244**, 42–3.

Galfrè, G. & Milstein, C. (1981). Preparation of monoclonal antibodies: strategies and procedures. In *Methods in Enzymology*, vol. 73, part B, eds. J. J. Langone & H. Van Vunakis, pp. 3–46. New York: Academic Press.

Galfrè, G., Milstein, C. & Wright, B. (1979). Rat × rat hybrid myelomas and a monoclonal anti-Fd portion of mouse IgG. *Nature*, **277**, 131–3.

Gefter, M. L., Margulies, D. H. & Scharff, M. D. (1977). A simple method of polyethylene glycol-promoted hybridization of mouse myeloma cells. *Somatic Cell Genetics*, **3**, 231–6.

Howard, J. C. (1985). Immunological help at last. *Nature*, **314**, 494–5.

Kabat, E. A. (1976). *Structural concepts in immunology and immunochemistry*, 2nd edition. New York: Holt, Rinchart & Winston.

Kearney, J. F., Radbruch, A., Liesegang, B. & Rajewsky, K. (1979). A new mouse myeloma cell line that has lost immunoglobulin expression but permits the construction of antibody-secreting hybrid cell lines. *Journal of Immunology*, **123**, 1548–50.

Kilmartin, J. V., Wright, B. & Milstein, C. (1982). Rat monoclonal antitubulin antibodies derived by using a new nonsecreting rat cell line. *Journal of Cell Biology*, **93**, 576–82.

Köhler, G. & Milstein, C. (1975). Continuous culture of fused cells secreting antibody of predefined specificity. *Nature*, **256**, 595–7.

Köhler, G., Howe, S. C. & Milstein, C. (1976). Fusion between immunoglobulin-secreting and non-secreting myeloma cell lines. *European Journal of Immunology*, **6**, 292–5.

Köhler, G., Pearson, T. & Milstein, C. (1977). Fusion of T and B cells. *Somatic Cell Genetics*, **3**, 303–12.

Leder, P. (1982). The genetics of antibody diversity. *Scientific American*, **246**, 72–83.

Littlefield, J. W. (1964). Selection of hybrids from mating of fibroblasts *in vitro* and their presumed recombinants. *Science*, **145**, 709–10.

Mitchison, N. A. (1971). The carrier effect in the secondary response to hapten-protein conjugates. I. Measurement of the effect with transferred cells and objections to the local environment hypothesis. *European Journal of Immunology*, **1**, 10–17.

Notton, B. A., Fido, R. J. & Galfrè, G. (1985). Monoclonal antibodies to a higher plant nitrate reductase: differential inhibition of enzyme activities. *Planta*, in press.

Porter, R. R. (1973). Structural studies of immunoglobulins. *Science*, **180**, 713–6.

Sanchez-Madrid, F., Szklut, P. & Springer, T. A. (1983). Stable hamster-mouse hybridomas producing IgG and IgM hamster monoclonal antibodies of defined specificity. *Journal of Immunology*, **130**, 309–12.

Secher, D. S. & Burke, D. C. (1980). A monoclonal antibody for large-scale purification of human leukocyte interferon. *Nature*, **285**, 446–50.

Sharon, J. (1980). Formation of hybridoma clones in soft agarose: effect of pH and of medium. *Somatic Cell Genetics*, **6**, 435–41.

Shulman, M., Wilde, C. D. & Köhler, G. (1978). A better cell line for making hybridomas secreting specific antibodies. *Nature*, **276**, 269–70.

Szybalska, E. H. & Szybalsky, W. (1962). Genetics of human cell lines. IV. DNA-mediated heritable transformation of a biochemical trait. *Proceedings of the National Academy of Sciences, U.S.A.*, **48**, 2026–34.

Trowbridge, I. (1978). Interspecies spleen-myeloma hybrid producing monoclonal antibodies against mouse lymphocyte surface glycoprotein, T200. *Journal of Experimental Medicine*, **148**, 313–23.

Williams, A. F. (1984). The T-lymphocyte antigen receptor – elusive no more. *Nature*, **308**, 108–9.

Williams, A. F., Galfrè, G. & Milstein, C. (1977). Analysis of cell surfaces by xenogeneic myeloma hybrid antibodies. Differentiation antigens on rat lymphocytes. *Cell*, **13**, 663–73.
Yarmush, M. L., Gates, F. T., Weisfogel, D. R. & Kindt, T. J. (1980). Identification and characterisation of rabbit-mouse hybridomas secreting rabbit immunoglobulin chains. *Proceedings of the National Academy of Sciences, U.S.A.*, **77**, 2899–903.

E. W. WEILER, J. EBERLE, R. MERTENS,
R. ATZORN, M. FEYERABEND,
P. S. JOURDAN, A. ARNSCHEIDT,
AND U. WIECZOREK

Antisera – and monoclonal antibody-based immunoassay of plant hormones

Plant hormone analysis – the problem

Nearly every aspect of the life cycle of a plant is controlled, or at least in some way influenced, by endogenous chemical regulators which are commonly termed "plant hormones" although in many respects these regulators do not conform to the definition of a "hormone" *sensu strictu*. These plant hormones are involved in the control of growth and differentiation as well as in the regulation of metabolic activities, e.g. in response to environmental stimuli. Displacement of the hormone balance – which may, for example be caused by pathogens – will quite often severely affect plant health, and mutations leading to alterations in hormone economy usually result in drastic changes in morphology, if not sublethal or lethal effects. Important members of the major groups of recognised plant hormones are shown in Fig. 1. While it is relatively easy to demonstrate the significance of plant hormones for the plant, it is less easy to detect hormones in the plant. However, a detailed knowledge of the levels, tissue and organ distribution, intracellular localization and translocation of plant hormones is a prerequisite before an understanding of how plant hormones work can be obtained. The old argument, as to whether plant hormones act through their levels, in a dose dependent manner or whether they remain a constant factor in an environment of changing hormone sensitivity have recently been reconsidered (Cleland, 1983, Trewavas, 1983) but in the absence of abundant and convincing data for either argument, it is more a matter of semantics at present. Our knowledge of plant hormones is scarce, indeed.

A thousand grams of a typical plant tissue will contain some 10 to 100 micrograms of most of the plant hormones at present of interest to the physiologist. Some – like the gibberellins (GAs) – may occur in quantities of less than a microgram. Cumbersome techniques of extraction and purification as well as sophisticated analytical devices (physico-chemical assays, bioassays) are usually required to estimate hormone levels in plants. Frequently, they yield information about a single hormone in a single plant species and with little or no spatial resolution. The delicate balance between several hormones acting at the same time, the subtle changes this balance may or may not undergo as a plant develops or responds to an environmental stimulus all too often cannot be elucidated due to the complex analytical problems involved.

Why use immunological techniques?

In this chapter the potential and limitations of the use of monoclonal antibodies (McAb) and antisera for the determination of plant hormones will be described. This is still a relatively novel approach in an area which is now dominated by the use of bioassay and various physico-chemical techniques (see Reeve & Crozier, 1980, for a recent review of these techniques). A word on the reasons which led us to explore the potential of immunoassay for plant hormone analysis may be appropriate.

Plant hormones can be analysed, by bioassay or physico-chemical assay only after extensive fractionation of plant extracts to remove impurities. The procedures required are frequently tedious and may lead to considerable

Fig. 1. Some endogenous plant growth regulators (plant hormones) of higher plants.

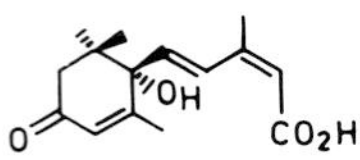

2-cis-(+)-Abscisic acid

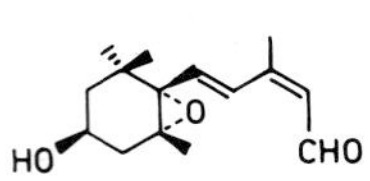

2-cis-Xanthoxin

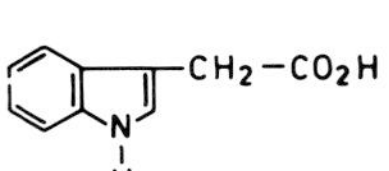

Indole-3-acetic acid

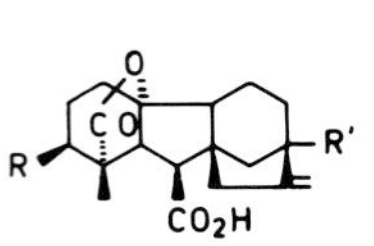

R = R' = OH Gibberellin A_1
R = OH, R' = H Gibberellin A_4
R = R' = H Gibberellin A_9
R = H, R' = OH Gibberellin A_{20}

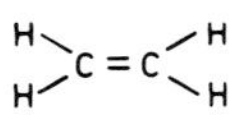

Ethylene

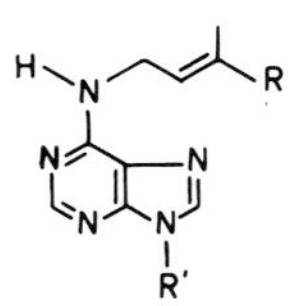

R = R' = H: Isopentenyladenine
R = H, R' = ribosyl: Isopentenyladenosine
R = H, R' = ribosyl-5'-P: Isopentenyladenosine-5'-Phosphate
R = OH, R' = H: trans-Zeatin
R = OH, R' = ribosyl: trans-Zeatinriboside
R = OH, R' = ribosyl-5'-P: trans-Zeatinribotide

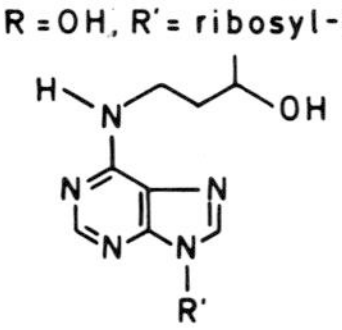

R' = H: Dihydrozeatin
R' = ribosyl: Dihydrozeatinriboside

losses of the hormone. Quantitative analysis of several hormones from small samples is often impossible and would, anyway, require extensive periods of time. Immunoassay (Berson & Yalow, 1959) exploits the intrinsic selectively of a high affinity antibody for its antigen. Binding is detected by the use of marker antigens labelled to very high specific activities. These assays have the potential to detect low amounts of plant hormones (by the binding process to their specific antibodies) in extracts of considerably lower purity than can be analysed by other techniques. This, plus the fact that immunoassays are convenient procedures, fast and relatively cheap, seemed attractive enough to us to embark on the development of immunoassay techniques for a range of key hormones such as indole-3-acetic acid (IAA), cis(+)abscisic acid (ABA), various gibberellins (GAs), cytokinins and others. It is now possible to quantify these hormones in femtomole amounts, sometimes even in crude or nearly crude extracts, with techniques so simple that they require a pipette, a plastic plate and an inexpensive colorimeter as sole equipment. With the development of McAb (see below), immunoassays should become widely accessible and they might well turn into routine procedures accompanying physiological, genetic, biochemical and molecular biological studies on plant hormones in the near future.

How to set up immunoassays

Advance considerations

Immunoassay exploits the reversible interaction of an antibody (Ab) with its antigen (Ag). By challenging a low concentration of Ab with the labelled tracer antigen (Ag*), again at low concentrations, the unknown concentration of Ag can be determined (see Fig. 2) from a reference dose-response curve constructed with known concentrations of standard Ag. Prerequisites of such assays are consequently the availability of a hormone specific antibody and a tracer hormone. Several options need to be considered with respect to the source and quality of antibody and the kind of label employed.

Fig. 2. Principle of competitive immunoassay.

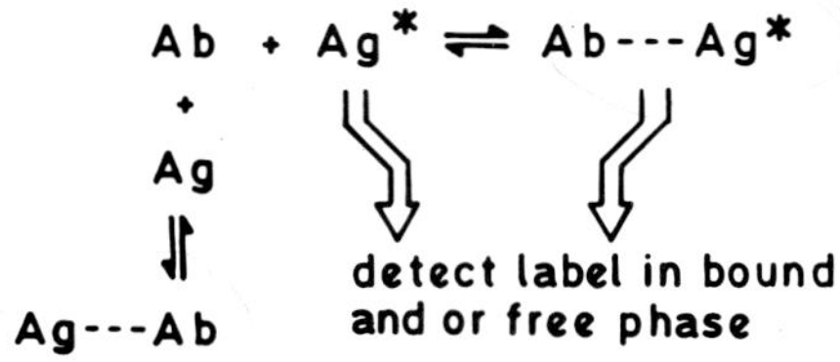

Monoclonal antibodies or antisera?

After the pioneering work of Köhler & Milstein (1975), McAb have rapidly evolved into one of the most powerful immunological tools. Using the techniques described (Galfrè and Butcher, this volume), as well as elsewhere (e.g. Goding, 1983), it is possible to generate unlimited quantities of pure McAb against any antigen to which the mouse or rat strain will respond initially. From a collection of McAb secreting hybridomas, those with just the right properties (e.g. affinity, specificity, stability) can usually be selected for further use. Further advantages of McAb are:

(a) that their production against specific antigens is possible even when the antigen cannot be obtained in pure form,
(b) that small amounts of antigen (sub-milligram) are sufficient as starting material,
(c) that contaminant antibodies – a major source of cross-reactivity – are, by definition, excluded, and
(d) that – because of the antibody uniformity – assay performance based on McAb is frequently better when using a highly modified (e.g. enzyme labelled) tracer antigen.

The drawbacks of McAb technology are equally remarkable: the generation of hybridomas is very time consuming, sometimes frustrating, definitely expensive and requires a specialised laboratory and immunological training. For the individual research laboratory without specialised immunology and cell culture facilities the production of antisera will be the faster, cheaper and safer way – even more so since we can already conclude from many comparisons between McAb and antisera that for the purpose of immunoassay alone, antisera are equally well suited. However, antisera-based assays are usually slightly less selective and antisera are frequently problematic when an adaptation for enzyme immunoassay (ELISA) is being attempted.

The selection of label

For a number of reasons (e.g. to maximize sensitivity, to trace highest-affinity antibodies in polyclonal mixtures), the tracer antigen concentration has to be kept to a minimum. Consequently, only labels of high specific activities are useful. This limits the choice to radioisotopes (radioimmunoasaay, RIA), enzymes (ELISA) and fluorochromes (fluoroimmunoassay, FIA). Due to the unavailability of cheap fluorescence analysers and the occurrence of highly fluorescent constituents in many plant extracts, FIA so far has attracted no attention in the plant hormone field. RIA and ELISA have been used extensively with ^{125}I and ^{3}H and alkaline phosphatase (an enzyme probably absent in plant material or present at low levels) as markers. The choice between a radioisotope and an enzyme will be influenced by the equipment and expertise available. In the long run, the use of enzymes is cheaper, safer, and requires less laboratory space and heavy equipment. ELISA is much simpler than RIA but to set up an ELISA using antisera can be very problematic. These problems arise since an antiserum may contain high affinity antibodies against the protein-hormone link in addition to those directed primarily against the hormone. These "bridge-binding" antibodies will not show up in most RIAs but will inevitably interfere if an enzyme-hormone conjugate contains the same, or a similar bridge as did the immunogen. It is a valid, but not absolute, generalization that RIA is to be preferred in combination with antisera and ELISA is the method of choice in combination with McAb.

Preparing the key reagents

Synthesis of immunogenic protein-hormone conjugates

All plant hormones are of low molecular weight and are not immunogenic *per se*. To render them immunogenic, they must be coupled as haptens, to a macromolecular carrier, preferably a protein. The hapten-concept was originally devised by Landsteiner (1945) and is an invaluable guide for those working on the immunological properties of small molecules.

From available evidence we can now conclude that, to a major extent, antibody specificity is governed by the structure of the conjugate, and not so much by the individual response of an animal – although this is clearly a

Table 1. *Comparison of specificity of antisera and McAb raised in different laboratories against trans-zeatinriboside.*

Compound	Type of antibody						
	Antisera						McAb
trans-zeatinriboside	100	100	100	100	100	100	100
trans-zeatin	44	60	34	42	41	46	47
cis-zeatinriboside	0.4	<1	0.8	nd	nd	1.4	0.8
dihydrozeatinriboside	1.7	33	2.1	1.3	nd	1.0	2.3
isopentenyladenine	nd	<1	0.2	nd	<1	0.4	0.9
isopentenyladenosine	0.1	<5	0.8	nd	<1.5	0.1	0.5
adenosine	0	<1	0	0	0.1	0	<0.002
	Weiler 1980b	MacDonald *et al.* 1981	Hansen *et al.* 1984	Badenoch-Jones *et al.* 1984	Ernst *et al.* 1983	Leroux 1984	Eberle *et al.* in press

Values are % cross-reactions on molar basis. nd = not determined.

source of variability. For example, cytokinin antisera raised in different laboratories against bovine serum albumin (BSA) conjugates of cytokinins synthesized according to the same protocol all are remarkably similar in their properties (Table 1). Even rabbit antisera and mouse McAb share these similarities (Table 1). As a consequence, in devising the synthetic route used for attaching the plant hormone to its protein carrier, one should have in mind the later application(s). Thus, if a hormone is conjugated through a functional group which is also used for conjugation by the plant, it is likely that endogenous hormone conjugates will cross react in the assay (which can be of advantage). If this is to be avoided, immunogen synthesis should ideally proceed through a position not attacked by plant enzymes. In practice, a compromise will frequently result; e.g. the coupling of indole-3-acetic acid through the indole nitrogen by a Mannich reaction as employed by Pengelly & Meins (1977) would be ideal, but results in very low titres, probably due to the chemical instability of the conjugated hormone. Coupling of IAA through the carboxyl function results in a stronger immune response – but some conjugates will be cross-reactive (see Table 3, below).

In Fig. 3, some routes for the synthesis of plant hormone-protein conjugates are summarized according to the functional groups involved. Serum albumins from bovine (BSA) or human (HSA) sources, or hemocyanin, were found useful as protein carrier. The molar ratios of hormone bound to protein in our experience does not influence the immunological response to a major extent. In some cases (e.g. for the cytokinins, GAs and ABA) ratios in the range of 4:1 to 10:1 (hormone:protein) have given satisfactory results while for IAA, higher ratios (15:1 to 30:1) do appear to work better. On the other hand, sera with high titres have been obtained against xanthoxin-BSA conjugates with coupling ratios of 1:1 (Feyerabend & Weiler, in prep.).

Radio- and enzyme-labelling of plant hormones

For the enzyme labelling of plant hormones, the same, or similar, synthetic routes may be used as for the synthesis of the immunogens (cf. Fig. 3). As mentioned already, special care as to the choice of the hormone-protein bridge may be required if ELISAs based on antisera are designed. There are a number of ways to avoid, or circumvent, the interference of bridge-binding effects. These include the following examples.

1. The use of a different coupling position at the hormone moiety.
2. The insertion of a chemically different bridge (e.g. by use of a spacer, or by omitting a spacer used in the immunogen synthesis). If carbodiimide is used to couple hormone to proteins, some

Fig. 3. Reaction sequences which were employed in the synthesis of plant hormone conjugates.

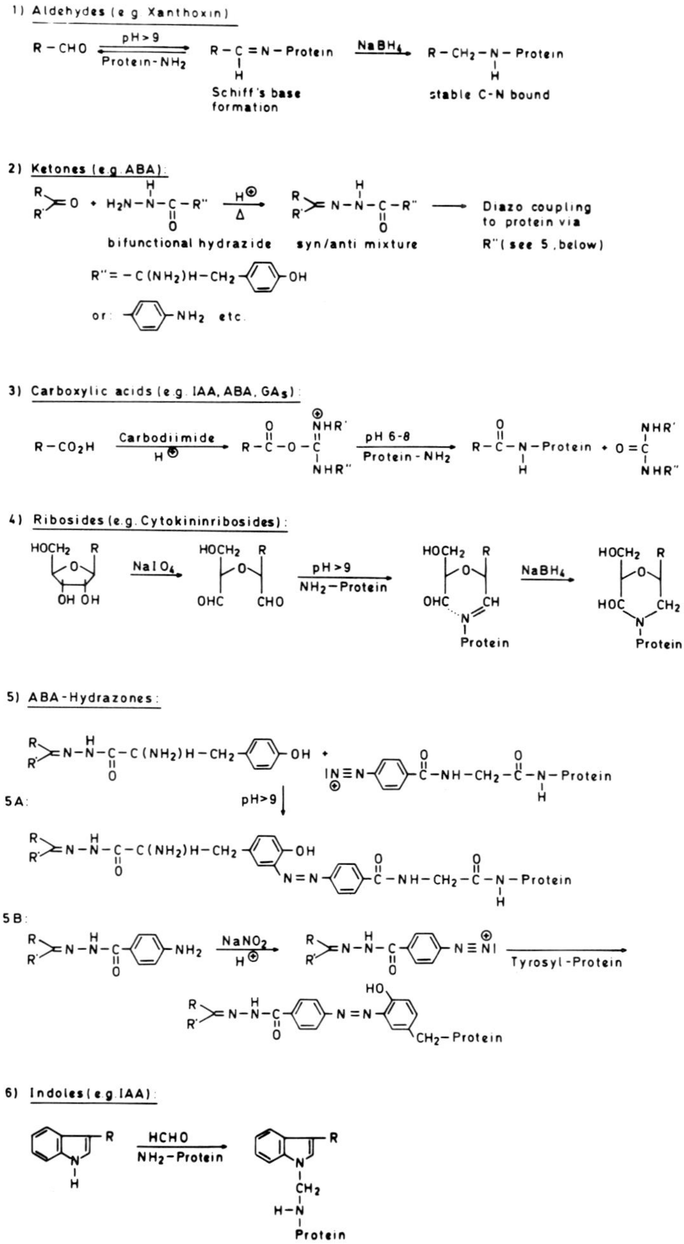

carbodiimide substitution of the protein may occur (see Erlanger, 1980). In this case, different carbodiimides should be used for immunogen and enzyme tracer synthesis.

3. Affinity purification of the antiserum (elimination of interfering antibodies with use of immobilized "bridges" or binding of specific antibodies to immobilized hormone – but here, see 2).
4. Chemical modification of the hormone. It has been a general observation that antisera and McAb raised against acidic plant hormones after coupling them via the carboxyl moiety, do not or only weakly bind the hormone in its anionic form. This is especially prominent for IAA and for the GAs. In these cases, methylation provides a good means of restoring immunoreactivity of these hormones – by removing the negative charge and inserting a small, but efficient "bridge"-analogue which adds to the overall affinity of the system.

Both ^{125}I and ^{3}H labelled plant hormones have been used for RIA. The structures of the tracers are shown in Fig. 4. Despite a considerably higher specific activity (ca. 2100 Ci $mmol^{-1}$ per ^{125}I), assay sensitivities with use of iodinated tracers are comparable or only slightly better than for assays employing tritiated tracer hormones. Tritium is more convenient to use because of its longer half-life and structural changes, if at all required, are less drastic than those introduced when incorporating iodine. Tritiated IAA, ABA and dihydrozeatin are now commercially available at high specific activities and tritiated GA methyl esters can be prepared conveniently as outlined in Fig. 4. Thus only the cytokinins, trans-zeatin and isopentenyladenine have to be used in a chemically modified form for RIA. The use of borohydride-reduced cytokinin riboside-dialdehydes (prepared by periodate treatment of the ribosides) results in stable tracers with satisfactory immunological properties.

Antibody production

Antisera

Antibody production has been treated in detail in the first chapter of this volume. A few remarks may suffice here to point out some observations made in our laboratory.

There are probably as many immunization schedules as workers in this field. Anyway, ours are given in Fig. 5. They have worked to our satisfaction in eliciting an immune response for a wide variety of hapten-protein conjugates and are easy to perform. Foot-pad injection of rabbits is not

Fig. 4. ^{125}I- and ^{3}H-labelled plant hormones and derivatives which were used in radioimmunoassay. A1: Condensation product of p-hydroxybenzaldehyde, putrescine and GA_3 as employed by Weiler and Wieczorek (1981); A2: Amide of IAA and tyrosine methyl ester (Weiler, 1981); A3: borohydride-reduced condensation product of tyramine and periodate-oxidized trans-zeatinriboside (Weiler, 1980b); I, (I): sites of possible iodine incorporation.

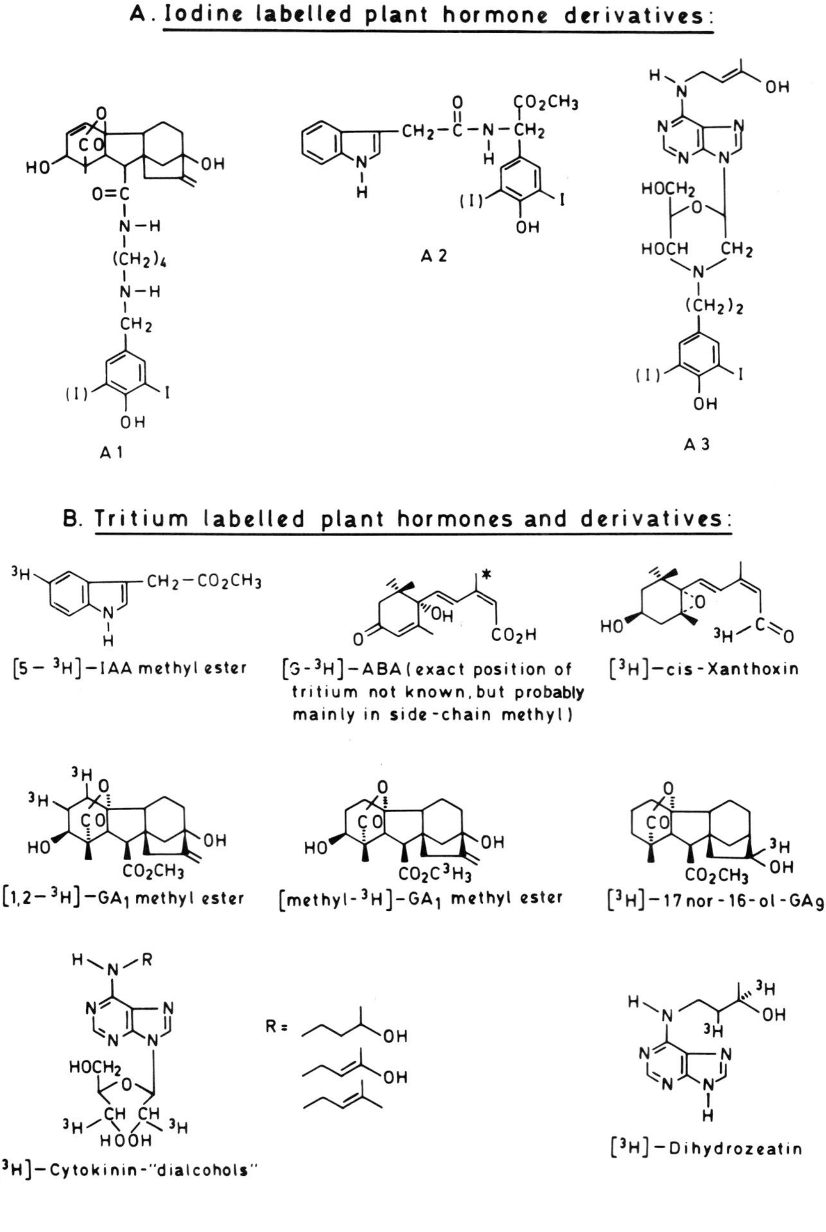

routinely used in our laboratory because of the stress imposed on the animals. However, this route is more efficient. In a recent comparison of immunization protocols for several GA-BSA conjugates (Yamaguchi *et al.*, unpublished) consistently higher titres have been obtained with the foot-pad injection as compared to the schedule given in Fig. 5.

It is essential to keep bleedings of the same animal and serum from different animals separate (store frozen at all times) until checked for titre and specificity. Premature pooling may ruin a valuable antiserum. For some hormone antigens, a higher inter-animal variability is observed than with others. For example, cytokinin conjugates will elicit a remarkably uniform response (cf. Table 1) whereas GA-BSA conjugates elicit a more variable response (S. Kurogochi *et al.*, unpublished)

Monoclonal antibodies

The general strategy for producing hybridomas in our laboratory is that developed by Galfrè & Butcher (this volume). We use P3-NS1-Ag4-1, X63-Ag8.653 and SP2/O-Ag-14 as myeloma lines and Balb/c mice as spleen donors. The immunization protocol given in Fig. 5 has so far been used successfully to immunize Balb/c mice with BSA conjugates of ABA, IAA and several cytokinins. It is essential to follow the development of serum titres because a higher yield of positive hybridomas can be expected from mice which show a strong immune response. Starting from such animals, a reasonable number of fusions (between 1 and 10) allowed the isolation of

Fig. 5. Immunization schedule routinely in use in our laboratory to immunize rabbits (left) and mice (right).

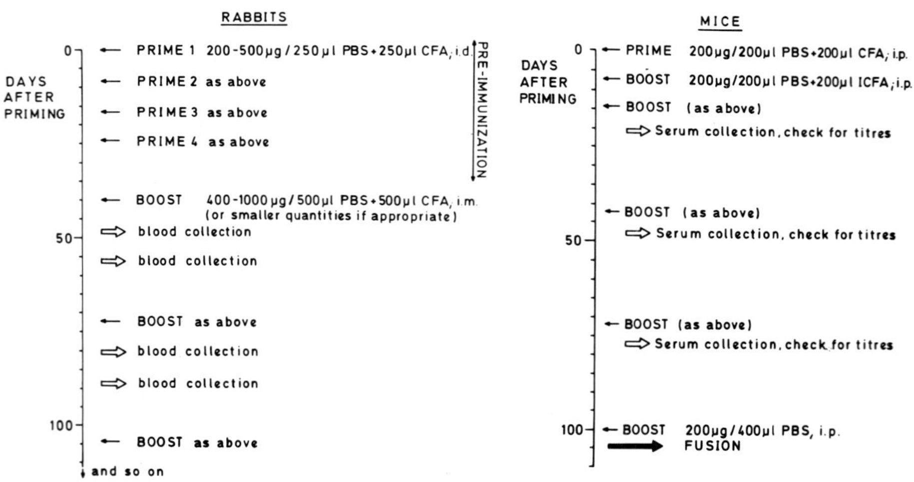

hybridomas secreting anti-hormone McAb with high affinity and specificity. RIA was employed as the screening procedure to ensure that only high affinity McAb were detected. Here, the concentration of tracer hormone is kept in the nM range and a positive binding indicates the presence of McAb with K_a values in that range. Positive results also confirm directly that the free hormone is being detected. If ELISA or dot blot assays are used as screening procedures, more positives are initially detected, but from these, the desired McAb have to be picked out by additional experiments.

The growth of hybridomas can be carried out *in vitro* (harvest of cell culture supernatants) or *in vivo* (ascites production). From cell culture supernatants, ammonium sulphate fractionation yields an McAb enriched fraction useful for ELISA and RIA; ascites fluid may be used directly. Further purification does not improve the assays. In addition, we make every attempt at mistreating our McAb from the beginning (storing cell culture supernatants frozen, checking repeated freezing and thawing as well as extended storage in solution at 4 °C). In this way, only stable McAb get into routine use – making life easier.

Bringing everything together: Assay optimization

We have now obtained an antiserum or a McAb after a long process of immunization and careful experimentation. We have also invested in synthesizing a radiolabelled or enzyme labelled hormone and we are now faced with the task of combining these reagents to devise an assay protocol which will ensure optimum assay performance. We like to use simple and uniform protocols for all our assays rather than having a multitude of different protocols to follow. Such generalized protocols are given in the Protocols section at the end of this chapter, for the RIA based on rabbit sera and for the ELISA based on mouse McAb. These protocols are general, in the sense that they work in our hands for all hormones we have tried so far. In the RIA, free and bound hormone are separated by ammonium sulphate precipitation of the immune complexes – a technique which is simple and reliable and not time dependent as is, for example, the dextran coated charcoal method which removes the free hormone leading to the dissociation of antigen-antibody complexes. The ammonium sulphate technique, however, is slightly temperature dependent and should not be carried out at temperatures higher than 25 °C (refrigerated centrifuge). The ELISA protocol recommended here uses a double layer of immobilized antibodies. As the first layer an antiserum rabbit anti-mouse Ig antibody preparation is used to saturate the polystyrene surfaces of microtitration plates. After washing, the specific McAb is bound to this antibody. At this stage, plates may be stored. This protocol has several advantages over the direct adsorption of the McAb to the plastic, *viz*:

1. Mouse McAbs tend to adsorb poorly to polystyrene in contrast to rabbit antisera.
2. The rabbit antibody acts as a buffer to compensate for plate-to-plate, batch-to-batch and even within-plate variability of adsorbed

Fig. 6. Principle of enzyme linked immunosorbent assays with immobilized antibody (above) and immobilized antigen (below) as the solid phases.

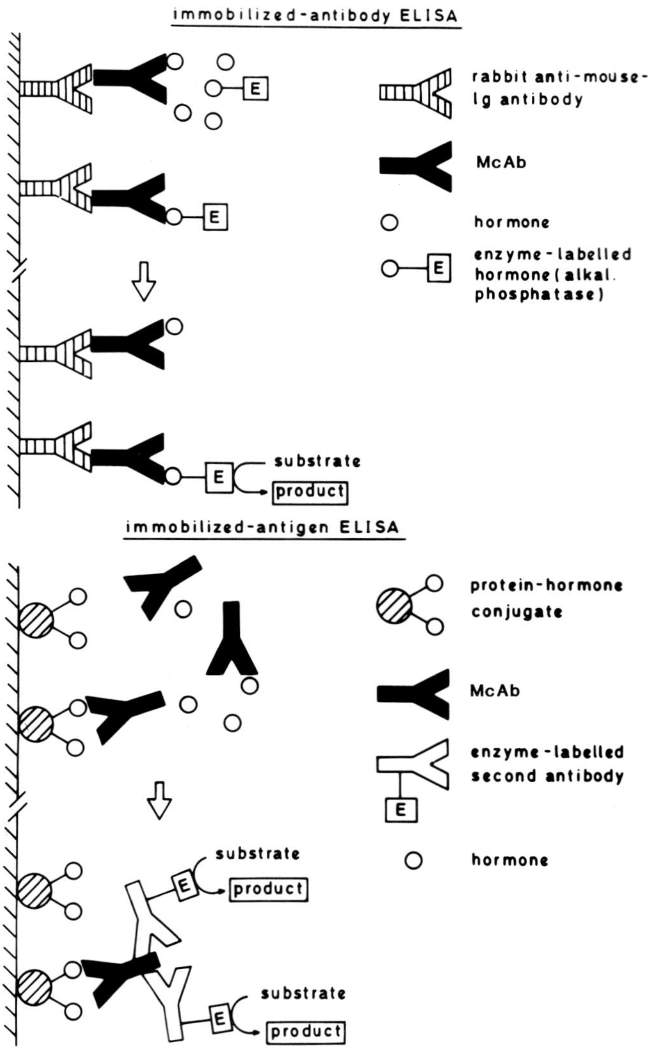

antibody. This variability is due to variations in the plastic composition and constitutes a considerable problem in "direct-coating" assays.

3. Unspecific binding of enzyme labelled hormone to the plastic surface is drastically reduced. The protocol as described produces coefficients of variation for triplicate standards of approximately 2% (within-plate variation).

In contrast to the immobilized-antibody technique detailed here, immobilized-antigen assays may be considered. The two quite different principles are shown in Fig. 6. It largely depends on the degree of unspecific binding and on the quality of adsorption of the hormone-protein conjugate used as immobilized antigen whether or not these techniques will be equally sensitive and reproducible as the immobilized antibody assays. Results in our laboratory indicate that immobilized-antigen assays are less sensitive and less precise than immobilized-antibody assays (J. Eberle, unpublished observation). General considerations concerning assay optimization are: (a) the choice of an appropriate amount of tracer hormone per assay, (b) the correct titration of serum/McAb to yield low assay variability and good sensitivity under conditions defined in (a), (c) the definition of the measuring range, checks for reproducibility of the assay, of the standard curve preparation, (d) definition of assay specificity, cross reaction analysis and (e) checks of long term stability of assay (components), effects of organic solvents on assay performance, etc. If at this stage, an assay works well, it is finally time to turn to the plant.

How to use immunoassays

Extraction and clean-up of plant material

Established procedures for plant hormone extraction and extract processing (see Yokota, Murofushi & Takahashi, 1980, for a survey) were devised to suit the needs of bioassay and physico-chemical assays. They may not be considered optimal for immuno-assays although they have been used extensively. An extract, or an extract fraction may be considered "immunologically pure" if (a) besides the hormone(s) to be analysed, no other cross-reacting material is present and if (b) it can be shown that the hormone-antibody interaction is not affected by other factors present in the sample. The processing scheme outlined in Fig. 7 involves solvent extraction, reversed-phase cartridge chromatography (recovery of all hormones considered here better than 85%) as well as partitioning and derivatization steps. It has given good results with a variety of extracts from different plants. The scheme must be checked for any new plant tissue under study before being considered "safe". For example, before

Fig. 7. General flow diagram for the processing of plant extracts prior to immunoanalysis. *Note:* this scheme requires validation for each plant material under investigation and modifications may be advantageous.

1. Freeze and lyophilize plant material, grind lyophilized sample

↓

2. Extract with 10 ml g^{-1} original fresh weight of 80 % MeOH containing 10 mg per liter of BHT at 4 - 10°C for 1 h (stir or use overhead mixer); add radioactive internal hormone standard at this stage if possible

↓

3. Pellet sample and decant, repeat step 2 two times; combine extracts

↓

4. Add water to a final MeOH concentration of 70 %

↓

5. Pass through reversed-phase C18-cartridge (SepPak, Bondelut etc.) to remove lipids and pigments

↓

6. Remove methanol with rotary evaporator or stream of nitrogen

check recovery of internal standard ⇦ **aqueous phase with plant hormones** ⇨ **McAb-ELISA for ABA**; **antiserum-RIA or McAb-ELISA for cytokinins**

↓

7. Adjust to pH 2.5 with 0.1 N HCl or suitable buffer

↓

8. Extract 3 times with equal vol. EtOAc, combine organic phase and concentrate to dryness

↓

9. Treat with excess of ethereal diazomethane (CAUTION) for 1-5 min on ice, destroy excess diazomethane with drop of 0.2 N HOAc in MeOH, dry down under nitrogen

↓

10. Redissolve in MeOH and dilute in assay buffer (final MeOH concentration not to exceed 10 %)

⇩ **McAb-ELISA for IAA methyl ester**; **antiserum-RIA for GA methyl esters**

⇩ check recovery of the internal standard(s)

step 7 is being carried out, neutral compounds interfering in the GA and/or IAA determination may have to be removed by partitioning first against ethyl acetate at pH 10. Cytokinins and GAs are thus determined as a group, dependent on the selectivity of the antibodies used (see the following section on antisera based assays). If the individual components are to be analysed, a suitable separation technique (e.g. TLC or HPLC) is to be used.

It is likely that future purification will involve the use of immobilized antibody for clean-up and concentration of hormones from crude extracts. A sequence of immobilized antibody columns could thus be used to retain specific hormones or groups of hormones. After elution from the columns the hormones are then analysed directly by immunoassay or are separated, for example, by HPLC and then analysed by immunoassay (Fig. 8). This scheme has several advantages: (a) most economic use is made of the sample and small samples could be processed as economically as large samples, (b) immunoaffinity columns can be used to concentrate traces of a hormone efficiently and (c) the technique provides a single-step clean-up useful for multiple-sample processing.

Fig. 8. Suggested flow diagram for the separation and isolation of plant hormones on immunoaffinity matrices.

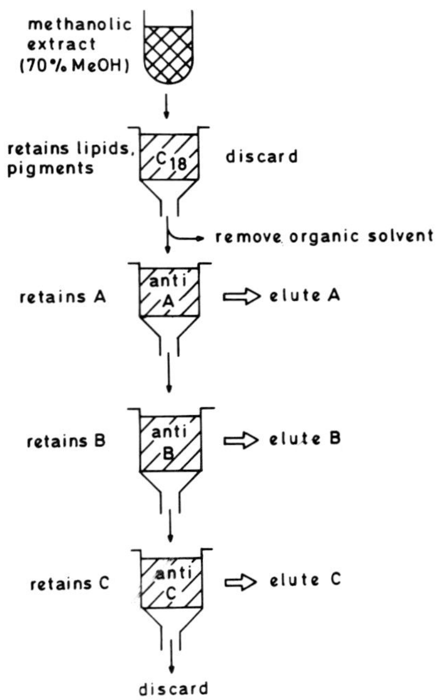

Akiyoshi *et al.* (1983), MacDonald & Morris (1983) and Zaerr, Durley & Morris (1983) have used immunoaffinity columns for the purification of cytokinins from several plant sources prior to analysis by HPLC and RIA. They report, however, on the instability of McAb when challenged with organic solvents. This is, however, not always a property of McAb. Mertens *et al.* (unpublished) have immobilized their McAb 15-I-C5 on CNBr-activated Sepharose and used this solid support for the single-pass immunoaffinity purification of ABA. As monitored by an internal standard of ^{3}H-ABA, the column retains ca. 85% of the hormone when applied in 50nM concentration in an aqueous plant extract fraction in a single pass. From the column, ABA is quantitatively eluted with acetone and the column stands more than 25 acetone cycles without wearing out. The further elaboration of the techniques for the immunoaffinity purification of plant hormones should considerably simplify hormone extract processing for immunoassay, bioassay as well as physico-chemical analysis.

Control of assay performance

Both assay precision, reproducibility and accuracy (degree of approximation to "true" value) need to be checked – as is the case with any analytical technique. The checks required reflect the sources of potential interference peculiar to immunoassays. These fall within two groups as below.

Fig. 9. Dilution analysis at several levels of sample added, in the presence of unlabelled, internal standard hormone. As an example (right), a dilution experiment for a sunflower hypocotyl extract after passage through SepPak C18 is shown. Analysis was performed with McAb-based RIA for ABA (clone 16-I-A4, Mertens *et al.*, 1983).

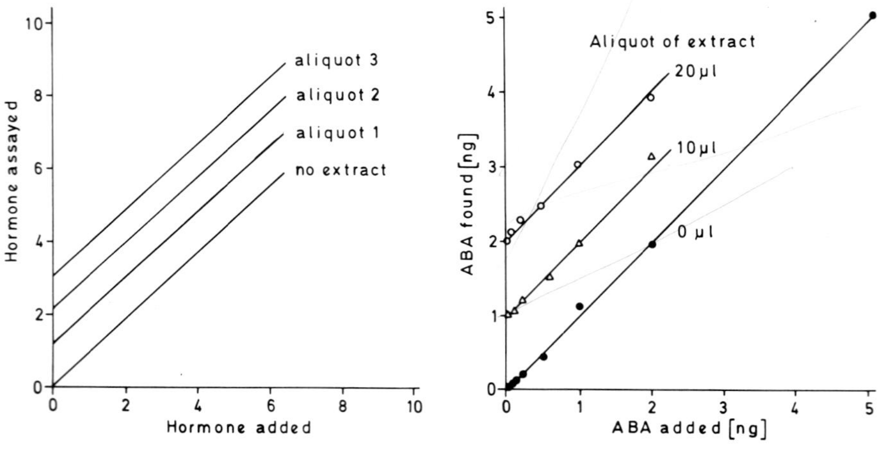

1. Compounds antigenically (structurally) similar to the plant hormone under study. These may exhibit high affinities for the antibody and thus will cross react even when present at low levels. Compounds with only a low cross reaction will be a problem if present in excessive amounts.
2. Factors interfering with certain steps of the assay procedure (e.g. detergents or solvent residues impairing antibody function, desorbing antibody from the solid phase etc., radioactivity quenchers in RIA if they co-precipitate with the immune complexes, inhibitors of the marker enzymes, etc.).

In addition, losses of hormone during extract work-up will affect accuracy. Work-up losses are conveniently detected by the use of radioactive hormone internal standards or by using hormone-spiked split extracts processed in parallel. The factors included in group (2) are best checked by a dilution analysis at various levels of added standard hormone as shown diagrammatically in Fig. 9 (left) and as an example, in Fig. 9 (right). Deviation from uniformity (slope = 1) indicates interference. In addition, cross reactants as defined under (1) will show up in this test if their dose-response curves do not run parallel to the hormone standard curve. Cross reactants with tracer displacement curves parallel to the standard curve cannot be detected by this method. Cross reactants are best detected in immunohistograms of separated extracts. An example is given in Fig. 10.

Ultimately, the validation of immunossay against physico-chemical techniques is desirable and if performed in conjunction with a "successive-approximation" approach as recommended by Reeve and Crozier (1980) yields solid data about assay accuracy. Available information suggests that immunoassays, when checked against GCMS, are accurate, provided the criteria set out above are fulfilled. This has been shown for a radio-immunoassay of IAA (Pengelly, Bandurski & Schulze, 1981), for several radioimmunoassays of cytokinins (Wang *et al.*, 1984, Ernst, Schäfer & Oesterhelt, 1983) and, using the successive approximation method and several HPLC purification steps, for the RIA of GA_9 (Atzorn & Weiler, 1983a) by P. Oden *et al.* (in preparation). Immunohistograms plus dilution analysis in the presence of hormone standard should be performed on any plant material under investigation before turning to the routine analysis of extracts or extract fractions.

The present state-of-the art

This chapter is intended to summarize information about available immunoassays for plant hormone analysis. Over the last few years, a considerable number of assays have been developed and put to use. These assays focus on the determination of key hormones considered to be of major physiological importance. However, the array of assays is by no means complete yet and a lot of work still needs to be done both to improve existing assays and to develop new ones.

Immunoassays based on antisera

Most available assays are based on antisera. These have been raised in rabbits against serum albumin and hemocyanin conjugates of the respective hormones and were used in RIA and, less frequently, ELISA (see Table 2, for a survey). The earlier complement fixation and T_4 bacteriophage inhibition assays are less sensitive and convenient. Both RIA and ELISA allow the quantitation of plant hormones at the femtomole level. These assays represent the most sensitive techniques known today and require less sample purification than other methods. Consequently, they have been used to estimate hormone levels in minute samples such as

Fig. 10. Immunohistogram of the extract for Fig. 9 (right), as analysed by McAb-based RIA (Mertens *et al.*, 1983) after HPLC separation on Lichrosorb RP8, 10μ (4.6 mm i.d. × 250 mm column size), flow rate 1.5 cm^3 min^{-1}. The closed bars represent the immunoreactive material, the open bars represent the radioactivity per fraction. As internal standard, ^3H-cis(±)ABA was used.

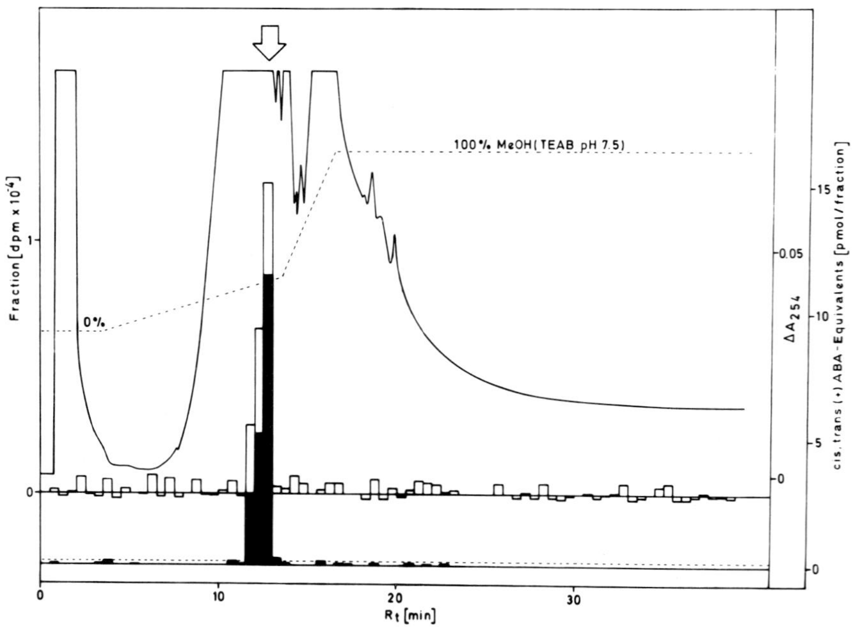

microlitre quantities of phloem sap (Weiler & Ziegler, 1981), guard cell protoplasts (Weiler, Schnabl & Hornberg, 1982), single coleoptiles (Weiler, Jourdan & Conrad, 1981), gravireacting coleoptiles, roots and hypocotyls (Mertens & Weiler, 1983), as well as for the determination of hormone gradients within plants (Hansen, Wenzler & Meins, 1984, Weiler *et al.*, 1981, Weiler, 1980a). Most of the existing RIAs employ tritium labelled hormones and the ammonium sulphate technique to separate free and bound hormone. However, Le Page-Degivry *et al.* (1984) employed equilibrium dialysis. The ELISA protocols typically involve polystyrene adsorbed antibodies and alkaline phosphatase as the marker enzyme. Leroux (1984) has successfully used the immobilized antigen technique and a biotinylated second antibody to detect antigen-bound specific antibody with an alkaline phosphatase labelled avidin. More information about the available assays can be derived from Table 2. It is not possible to discuss here in detail the specificities of the various antibodies which are in use. In general, cross-reactions that were observed are restricted to extremely similar compounds (see Table 2). Cross reactions between different groups of hormones have not been reported and are not expected because of the structural differences between these compounds.

Monoclonal antibody-based assays

Very recently, Trione & Morris (1983) as well as Mertens, Deus-Neumann & Weiler (1983) have succeeded in raising McAb against plant hormones. The general advantages of McAb over antisera have been discussed already. However, in order to be useful for immunoasaay, both a high degree of specificity and a high affinity are required. In our laboratory, we have selected for these characteristics by employing RIA as screening systems (for reasons, see above). It is clear now that properly selected McAb are at least competitive to antisera in terms of specificity (Mertens *et al.*, 1983, Table 1, Table 3). The major characteristics of the hybridoma lines raised so far in our laboratory and of the secreted McAb are given in Table 4. The use of these McAb has allowed the establishment of both RIA and ELISA procedures which all follow the protocols given in part 8 of this chapter. Especially attractive are the ELISA techniques which exhibit far better precision (insignificant unspecific binding, steeper slope of standard curves) and reproducibility (coefficients of variation for replicates) than antisera-based ELISAs. The important parameters of these assays are also included in Table 4 and Figure 11 shows a typical computer printout of a standard curve for the ELISA of trans-zeatinriboside.

The major drawback of McAb-based ELISA of plant hormones at present is that techniques are available for only a limited range of compounds.

Table 2. *Survey of plant hormone immunoassays based on antisera.*

Abbreviations used:
ABA – abscisic acid; AC – aminocaproic acid; AP – alkaline phosphatase; 6-BAP – 6-benzylaminopurine; BSA – bovine serum albumin; CFI – complement fixation inhibition; DHZ – dihydrozeatin; ELISA – enzyme linked immunosorbent assay; GA – gibberellin; GLC glucoside; GLE – glucosyl ester; HCY – hemocyanin; HSA – human serum albumin; IAA – indole-3-acetic acid; IBS – indole-3-butyric acid; IP – isopentenyladenine; IPA – isopentenyladenosine; JA – jasmonic acid; ME – methyl ester; MT – methylthio-; 1-NAA – 1-naphthylacetic acid; -P – phosphate; PHI – T4-phage inhibition assay; RIA – radioimmunoassay; tZ – trans-zeatin; tZR – trans-zeatinriboside.

Immunogen	Coupling site	Hormone standard	Assay system	Detection limit (pmol)[a]	Assay quantitates	Reported cross reactants[b]	Reference
I. Indole-3-acetic acid:							
IAA-BSA	-N	IAA	RIA/^{3}H	1	IAA	1-NAA	Pengelly & Meins, 1977
IAA-BSA	-COOH	IAA	ELISA/AP	0.2	IAA	1-NAA	
						IAA-L-alanine	
						indolepyruvate	Leroux, 1984
IAA-HCY	-COOH	IAA	CFI	300/ml	IAA	not given	Fuchs & Fuchs, 1969
IAA-HCY	-COOH	IAA-AC	PHI	60/ml	IAA	not given	Fuchs *et al.*, 1971
IAA-BSA	-COOH	IAA-ME	RIA/^{3}H	0.5	IAA-ME	1-NAA-ME	
						indoleacetaldehyde	
						indoleacetamide	Weiler, 1981
IAA-BSA	-COOH	IAA-ME	RIA/^{125}I	1	IAA-ME	1-NAA-ME	
						indoleacetaldehyde	
						IAA-aspartate	
						indoleacetamide	Weiler, 1981
IAA-BSA	-COOH	IAA-ME	ELISA/AP	0.02	IAA-ME	IBS-ME	
						indoleacetamide	
						IAA-aspartate	
						1-NAA-ME	Weiler *et al.*, 1981
II. Abscisic acid:							
(±)ABA-HCY	-COOH	(±)ABA	PHI	40/ml	(±)ABA	not given	Fuchs *et al.*, 1972
cis(±)ABA-BSA	-COOH	cis(±)ABA	RIA/^{3}H	0.25	cis(±)ABA	ABA-ME	
						ABA-$(OAc)_4$-GLE	Weiler, 1979
cis(±)ABA-BSA	-COOH	cis(±)ABA	RIA/^{3}H	0.40	cis(±)ABA	ABA-ME	
						ABA-$(OAc)_4$-GLE	Walton *et al.*, 1979
cis(+)ABA-BSA	-COOH	cis(+)ABA	RIA/^{3}H	0.20	cis(+)ABA	ABA-ME	
						ABA-GLE	Weiler, 1980*a*
cis(±)ABA-BSA	-C=O	cis(+)ABA	RIA/^{3}H	0.10	cis(+)ABA	trans-ABA	Weiler, 1980*a*
cis(±)ABA-BSA	-COOH	cis(±)ABA	ELISA/AP	0.10	cis(±)ABA	ABA-ME	Daie & Wyse, 1982
cis(+)ABA-BSA	-COOH	cis(+)ABA	ELISA/AP	0.06	cis(+)ABA	ABA-ME	

cis(±)ABA-HSA	-COOH	cis(±)ABA	RIA/[illegible]I	0.20	cis(±)ABA	ABA-ME, ABA-GLE, ABA amides	LePage-Degivry *et al.*, 1984
cis(±)ABA-HSA	-COOH	cis(±)ABA	RIA/^{3}H	0.28	cis(±)ABA	not given	Kannangara *et al.*, 1984
cis(+)ABA-BSA	-COOH	cis(+)ABA-ME	ELISA/AP	0.02	cis(+)ABA-ME	not given	Leroux, 1984
cis(±)ABA-BSA	-COOH	cis(±)ABA	RIA/^{3}H	1	cis(±)ABA	not given	Southwick *et al.*, 1984
III. Cytokinins							
IPA-BSA	-ribosyl	IPA	RIA/^{3}H	10	IPA	not given	Khan *et al.*, 1977
IPA-BSA	-ribosyl	IPA	RIA/^{3}H	0.12	IP, IPA, IPA-5′-P	6-BAP, 6-BAP-9-GLC	Weiler & Spanier, 1981
IPA-BSA	-ribosyl	IPA	RIA/^{125}I	4/ml	IP, IPA	6-BAP	deGreef *et al.*, 1980
IPA-BSA	-ribosyl	IPA	RIA/^{3}H	1	IP, IPA, IPA-5′-P	2-MT-IPA	MacDonald *et al.*, 1981
IPA-BSA	-ribosyl	IPA	RIA/^{3}H	0.5	IP, IPA	6-BAP	Ernst *et al.*, 1983
IPA-BSA	-ribosyl	IPA	RIA/^{3}H	0.5	IP, IPA	6-BAP	Cuttin *et al.*, 1984
tZR-BSA	-ribosyl	tZR	RIA/^{3}H	0.04	tZ, tZR, tZR-5′-P	2-MT-tZR	Weiler, 1980b
tZR-BSA	-ribosyl	tZR	RIA/^{125}I	0.14	tZ, tZR	—	Weiler, 1980b
tZR-BSA	-ribosyl	tZR	RIA/^{3}H	10	tZR	2-MT-tZR	Vold & Leonard, 1981
tZR-BSA	-ribosyl	tZR	RIA/^{3}H	0.04	tZ, tZR	DHZ	MacDonald *et al.*, 1981
tZR-BSA	-ribosyl	tZR	RIA/^{3}H	0.3	tZ, tZR, tZ-9-GLC	9-alanyl-tZR	Badenoch-Jones *et al.*, 1984
tZR-BSA	-ribosyl	tZR	ELISA/AP	0.3	tZ, tZR	—	Hansen *et al.*, 1984
IV. Gibberellins							
GA3-HCY	-COOH	GA3	CFI	10/ml	GA3	not given	Fuchs & Fuchs, 1969
GA3-HCY	-COOH	GA3	PHI	10/ml	GA3, GA1	GA7, GA4	Fuchs *et al.*, 1971
GA3-BSA	-COOH	GA3-ME	RIA/^{125}I	0.005	GA3-ME	GA1,GA7-ME	Weiler & Wieczorek, 1981
GA1-BSA	-COOH	GA1-ME	RIA/^{3}H	0.07	GA1-, GA4-	GÀ7-, GA20-ME	Atzorn & Weiler, 1983*a*
GA3-BSA	-COOH	GA3-ME	RIA/^{3}H	0.1	GA3-ME	GA7-, GA20-ME	Atzorn & Weiler, 1983*a*
GA9-BSA	-COOH	GA9-ME	RIA/^{3}H	0.1	GA9-ME		Atzorn & Weiler, 1983*a*
GA3-BSA	-COOH	GA3-ME	ELISA/AP	0.0005	GA3-, GA7-ME	GA20-ME	Atzorn & Weiler, 1983*b*
GA4/7-BSA	-COOH	GA4-, GA7-ME	ELISA/AP	0.001	GA4-, GA7-ME	GA32-ME	Atzorn & Weiler, 1983*b*
GA24-BSA	-C=CH2	GA24-ME or GA19-ME	RIA/^{3}H	0.2	GA24-, GA19-ME	GA12-, GA20-ME	Kurogochi *et al.*, in prep.
V. Other compounds							
Xanthoxin-BSA	-CHO	xanthoxin	RIA/^{3}H	1	xanthoxin	violaxanthin, oth. xanthophylls	Feyerabend & Weiler, in prep.
JA-BSA	-COOH	JA-ME	RIA/^{3}H	10	JA-ME	dihydro-JA	Knöfel *et al.*, 1984

Note: [a]Data were estimated from standard curve data if values were not available from text; [b]Data were compiled from available information in the references, this does not mean that compounds not mentioned here are not immunoreactive.

Table 3. *Comparison of cross reactions of an antiserum and a monoclonal antibody directed against IAA, both raised in our laboratory.*

All compounds were treated with diazomethane prior to analysis.

Compound	% Cross reaction on molar basis McAb 17-II-A1	Antiserum[a]
Indole-3-acetic acid	100	100
Indole-3-acetone	5.2	35
Indole-3-propionic acid	0.5	2.5
Indole-3-butyric acid	1.4	10.2
Indole-3-acetonitrile	1.0	2.8
Indole-3-pyruvic acid	0.1	1.5
Indole-3-lactic acid	0.5	0.5
Indole-3-acrylic acid	7.0	0.5
Indole-3-ethanol	0.04	1.6
4-Cl-indole-3-acetic acid	0.9	91
5-OH-indole-3-acetic acid	0.02	0.3
6-OH-indole-3-acetic acid	0.2	0.9
Indole-3-acetamide	1.0	8.6
Indole-3-acetylglycine	58	18
Indole-3-acetyl-L-alanine	1.5	3.1
Indole-3-acetyl-L-phenylalanine	0.6	11.3
Indole-3-acetyl-D,L-aspartic acid	0.5	2.9
1-Naphthylacetic acid	0.1	6.7
Phenylacetic acid	0.01	0
2,4-Dichlorophenoxyacetic acid	0.01	0

Note: [a]Weiler *et al.* (1981), and new data.

Table 4. *Properties of some monoclonal antibodies to plant hormones and of ELISA systems based on these antibodies.*

Abbreviations as in Table 2. DHZR = dihydrozeatinriboside.

Hormone standard	Clone	Myeloma line	Ig-class	Affinity (l/M)	Assay system	Measuring range (pmol)
cis(+)ABA	16-I-A4	NS1	A	7.9×10^9		
	15-I-C5	NS1	G1	2.6×10^8	ELISA/AP	0.02 – 2
IAA-ME	17-II-B3	NS1	G2a	4.6×10^7		
	17-II-A1	NS1	G2a	1×10^8	ELISA/AP	0.2 – 20
	18-III-C3	NS1	G2b	3.2×10^7		
tZR	J3-I-B3	NS1	G1	1.1×10^8	ELISA/AP	0.02 – 5
	J18-I-D1	X63.Ag8	G2b	1.6×10^8		
	J18-III-B5	X63.Ag8	G1	9.1×10^7		
DHZR	J23-II-B1	NS1	G1	2.3×10^8	ELISA/AP	0.1 – 3

It is to be expected, however, that further assays will be forthcoming soon. These techniques may eventually replace antisera-based assays to a large extent.

Mertens *et al.* (1983) have reported that their McAb against ABA has lower cross reactivities than antisera raised against the same hapten. The same was observed for McAb against IAA (Mertens *et al.*, 1985, see Table 4) although there are exceptions. Eberle *et al.* (1986) have generally found comparable specificities of antisera and McAb directed against cytokinins (for an example, see Table 1). It can be concluded, then, that McAb are comparable to, if not better than, antisera as far as specificity is concerned. Mertens *et al.* (1983) did not detect cross reacting materials interfering with their McAb-based RIA for cis(+)ABA when analysing crude methanolic leaf extracts of *Vicia faba*. C_{18}-cartridge purified hypocotyl extracts of light grown sunflower seedlings gave similar results (Fig. 10). The acidic ethyl acetate fraction of maize coleoptile extracts (after passage through a C_{18}-cartridge and EtOAc extraction at pH 11) was practically devoid of immunoreactive material other than IAA when tested in a McAb-based ELISA for IAA methyl ester (Mertens *et al.*, 1985).

HPLC-immunoassay

The combination of HPLC and immunoassay is especially attractive for solving more complex analytical problems. HPLC prior to immunoassay may be useful to (a) remove interfering compounds from extracts if less

Fig. 11. Typical standard curve for the ELISA of trans-zeatinriboside with McAb J3-I-B3. Shown is a computer-generated standard curve using a polynomial regression algorithm to calculate the curve. The crosses mark the data points as measured (triplicates).

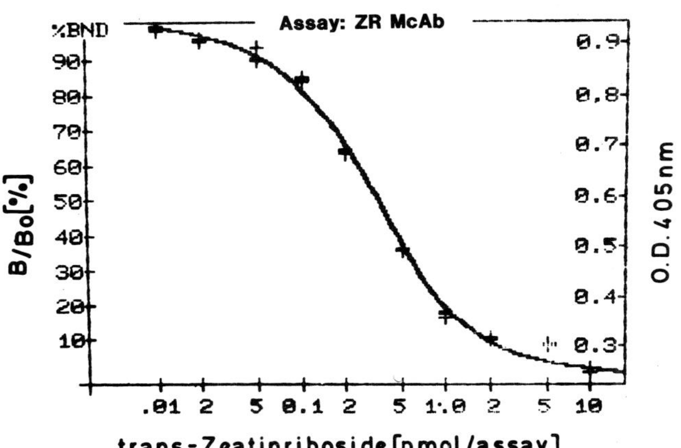

selective antisera are available, (b) separate immunologically indistinguishable components of a hormone mixture (e.g. cytokinin ribosides and free bases), (c) provide group separation of hormones prior to immunoassay, (d) monitor the purity of immunoaffinity purified hormone(s) and (e) evaluate results obtained by immunoassay.

In contrast to other chromatographic techniques (e.g. gas chromatography), HPLC, like TLC, is applicable to all plant hormones (if we exclude ethylene) without derivatization. TLC has a weaker separation power and the further disadvantage that sample losses on the dried layers can be severe. In contrast, HPLC allows the recovery of plant hormones in good yield, if certain precautions are observed. HPLC in combination with immunoassay has been used for the analysis of cytokinins (e.g. MacDonald, Akiyoshi & Morris, 1981, Regier & Morris, 1982), ABA (Kannangara *et al.*, 1984, see also Fig. 10) and IAA (Mertens *et al.*, 1985, Schröder *et al.*, 1984). For cytokinin analysis, HPLC can be used efficiently to separate the various 9-substituted analogues of the cytokinin bases. All antibodies available sharply discriminate the bases but cross-react with a range of 9-substituted derivatives. This can be advantageous since it allows the group quantification of a series of biogenetically closely related cytokinins with a single immunoassay.

In addition, the use of group specific antibodies in conjunction with HPLC appears to be an effective solution to the immunoassay of GAs. This area has not been explored in detail so far. Antisera which react with a series of closely related GAs are already available (Atzorn & Weiler, 1983*a*, *b*, S. Kurogochi *et al.*, in preparation, J. Eberle *et al.*, in preparation). A systematic study, however, is still lacking.

A look into the future of plant hormone immunoassay

Immunoassays are an emerging technique for plant hormone detection now gradually becoming acquired by an increasing number of laboratories. In fact, immunoassay is such a flexible technique that there seems little limitation, other than the imagination of the researcher, for the improvement of existing assays and the development and use of new ones. We want to conclude this review with a personal, and in a way subjective, glimpse on likely future developments in this field which might stimulate those interested in adopting the technique for their own needs.

1. There can be little doubt that although more time consuming and costly, McAb will probably become increasingly important in the assay of plant hormones. Whenever a rare compound is not available in pure form, or in insignificant quantities, McAb will be the preferred approach. With modern techniques of *in vitro*

immunization, microgram quantities of hormone-protein conjugates might allow the production of McAb secreting hybridomas. Furthermore, the increased specificities of McAb will allow a further reduction of extract processing steps. McAb can be produced in any desired quantity thus allowing their widespread distribution and the development of preparative systems for hormone purification and of techniques for hormone immunolocalization.

2. Immunoassays will become increasingly simple. Whereas RIA requires substantial equipment and is a multi-step procedure, ELISA requires little expensive equipment and involves fewer assay steps. The use of McAb allows for uniform assay protocols and reduces assay optimization time to a minimum. Such assays consequently are easy to adopt, to execute in parallel and they can be performed by automated analysers which are already available in the clinical market. Future simplifications in assay performance might involve homogenous enzyme immunoassays (e.g. Rubenstein, Schneider & Ullman, 1972), i.e. techniques which require no separation of free and bound hormone such as fluorescence excitation transfer assays (Ullman, Schwarzenberg & Rubenstein, 1976) or enzyme inhibition assays (Rubenstein *et al.*, 1972). Such assays, however, are still less common than ELISAs which require little effort to separate free and bound hormone.
3. The range of assays for different hormones will be extended. Whereas the 'first' generation of assays was developed against the key hormones considered physiologically active, future assays will include those for immediate or distant biosynthetic precursors, hormone catabolites and hormone conjugates. In addition, esoteric growth regulators and synthetic ones will become accessible to analysis. Certain enzymes involved in hormone biosynthesis can be studied with antibodies specific for their substrates or products. Such antibodies would allow the development of assays which would facilitate the purification of the enzymes and the elucidation of their properties. Immunoassay can be used as very sensitive assay for enzymatic activity if spectrophotometric or radioassays are unavailable. For example, RIA has been used by Treimer and Zenk (1978) to study the conversion of tryptamine and secologanine to *Corynanthe*-type alkaloids whilst the function of the gene 2 product of *Agrobacterium tumefaciens* T-DNA became apparent in a screening programme which employed the ELISA for IAA (Weiler *et al.*, 1981) to detect hormone synthesized in cell free

systems containing the gene 2 product and various potential precursors of IAA (Schröder *et al.*, 1984). The extension of this approach to biosynthetic pathways of other hormones should be most promising.

4. The exploitation of a combination of efficient and fast separation techniques, such as HPLC, with group selective antisera should facilitate the analysis of complex mixtures of hormones (e.g. GAs) using a limited number of assays.
5. The potential to immunoanalyse dozens of plant extracts per day for several hormones will certainly create pressure to re-evaluate hormone extraction and simplify extract processing. The use of appropriate combinations of immunoaffinity columns and precolumns (such as e.g. C_{18}-cartridges) should allow a fast clean-up of samples and to reach hormone abundances in the eluted fractions suitable for direct analysis by physico-chemical techniques, e.g. by GCMS. The combination of optimized tissue extraction and extract processing with immunoassay might well render the analysis of many hormones routine within the next few years.
6. Finally, the availability of hormone specific antisera and of McAb will trigger the use of these for other purposes. We have already mentioned the inclusion of immunological steps in advanced hormone purification protocols. This does not only relate to the isolation of hormones from plant sources but could also include the facile purification of, for example, radiolabelled hormones. Enantiomer separation has been achieved by Mertens *et al.* (1982) who purified tritiated cis(+)- and cis(−)-ABA from racemic mixtures using immunoaffinity matrices.

In addition, the immunocytochemistry of plant hormones might now come into focus to gain insights into tissue and organ distributions of plant hormones. First reports using antisera have already appeared (Zavala & Brandon, 1983, Sotta *et al.*, 1985). The difficulties which need to be overcome here are partly technical in nature (appropriate immobilization of hormone, avoidance of sectioning artifacts etc.) and partly theoretical in nature (e.g. devising proper controls allowing one to interpret results unequivocally). However, the localization of small molecules such as serotonin in tissue sections has become possible in animal tissues (e.g. Nemoto *et al.*, 1983) and this is definitely encouraging.

In conclusion, then, we feel rather optimistic that the further use of antibodies and immunoassay will help to solve at least some of the many problems still associated with understanding what plant hormones actually do. The technical difficulties in analysing plant hormones might have discouraged many to enter this area. We hope that with an increasing use of immunological reagents for plant hormones, this situation will eventually change.

PROTOCOLS

I. Radioimmunoassay of plant hormones using rabbit antisera

1. Dispense 0.1 cm^3 of dilute hormone standard or sample into test tubes[a].
2. Add 0.7 cm^3 of RIA incubation mix to each tube and vortex gently[b].
3. Add 0.1 cm^3 of dilute antiserum[c] or 0.1 cm^3 of PBS[d].
4. Mix the contents of the tubes by vortexing gently.
5. Incubate for 90 min in the cold (4–10 °C)[e].
6. Add 1 cm^3 of saturated ammonium sulphate solution diluted 10 + 1 with water pH 7.4 and incubate in the cold for 30 min.
7. Centrifuge at ca. 2000 × g for 10 min.
8. Decant and discard supernatants.
9. Wash the pellets once with 1 cm^3 half saturated ammonium sulphate solution (equal volumes of saturated ammonium sulphate and water) and re-centrifuge.
10. Decant thoroughly, leaving for at least 5 min on soft, absorbant tissue to remove as much supernatant as possible.
11. Dissolve pellets in 0.3 cm^3 water.
12. Add 1.5 cm^3 Quickszint 212 and mix well until clear one-phase system results[f].
13. Calculate dose-response curves (logit/log plot) for standards and logit B/B_0 for the samples[g].

II. Enzyme immunoassay of plant hormones based on mouse monoclonal antibodies

1. Coat polystyrene microtitration plates (96-well or similar) overnight at 4–10 °C with 0.2 cm^3 of RAMIG[h].
2. Decant thoroughly, add 0.2 cm^3 of dilute McAb and incubate for 24 h at 4–10 °C[i].
3. Decant and rinse twice with tap water.
4. Add, to each cup, 0.05 cm^3 TBS, 0.1 cm^3 of hormone standard or sample in TBS[j].
5. Incubate plates for 60 min at 4–10 °C[k].
6. Add 0.05 cm^3 of dilute enzyme tracer, mix and incubate for 3 h at 4–10 °C[l].
7. Decant and rinse plates twice with tap water.
8. Add 0.2 cm^3 of a fresh solution of 1 mg p-nitrophenyl phosphate (e.g. Sigma 104) in 1 cm^3 of 50 mM $NaHCO_3$, pH 9.6.
9. Stop reaction with 0.05 cm^3 of 5 M KOH.
10. Read absorbances at 405 nm[m].

Notes

a Dilute hormone standards from methanolic stock solution (e.g. 0.1 mg cm^{-3}) with assay buffer or water. (We use 12 × 63 mm glass tubes which fit into standard scintillation vials and into most counters.) Run all standards and samples in triplicate. Use buffer instead of hormone standard for B_0 (B_{max}) values.

b Prepare sufficient RIA incubation mix: 5 parts of PBS (prepared by dissolving 0.15 M NaCl in 0.01 M phosphate, pH 7.4), 1 part of dilute bovine serum (prepared by diluting bovine serum 6–8 fold with distilled water (final protein concentration approximately 0.7 mg cm^{-3} as assayed with the Biorad Kit)), and 1 part of aqueous radiotracer dilution to give desired activity (approximately 20 nCi per tube).

c Dilute antiserum from stock or intermediate dilution appropriately with PBS, keep on ice, use the same day. Store serum stock or tenfold diluted (PBS) stock frozen in small aliquots. Note: repeated freezing and thawing might result in loss of activity. (Dilution chosen to bind 30–35% of the radiotracer present under equilibrium conditions; if high sensitivity is not required and precision is a problem, use enough antiserum to bind 50% of the tracer.)

d For determination of unspecific binding (UB values).

e Incubate in the darkness when assaying for IAA or ABA.

f Note: other scintillation cocktails will require different cocktail to water ratios. Determine radioactivity in a scintillation counter using counts preset if time is not a problem.

g Relative binding (B/B_0) = ($cpm_{sample} - cpm_{ub}$)/($cpm\ B_0 - cpm_{ub}$) and logit (B/B_0) = (1n (B/B_0)/(100 – B/B_0) where B_0 = maximum binding in the absence of standard or sample and ub = unspecific binding in the absence of antibody. Logit (B/B_0) vs. log antigen plots should be linear in the desired measuring range.

h Dilute appropriate amount of lyophilized rabbit anti-mouse immunoglobulin (RAMIG) in 50 mM $NaHCO_3$ buffer, pH 9.6. We prepare our RAMIG from rabbits immunized with mouse immunoglobulin. The immunoglobulins are precipitated from the rabbit sera by the rivanol/ammonium sulphate technique (see Hurn and Chantler, 1980), dialysed and lyophilized. This preparation is then dissolved in 50 mM $NaHCO_3$ buffer typically at about 0.25 mg cm^{-3}.

i We dissolve the appropriate amount of McAb fraction (lyophilized ammonium sulphate fraction of cell culture supernatants) in 50 mM $NaHCO_3$ buffer or we dilute ascites fluid in a TBS-gelatine solution. TBS is prepared by dissolving 150 mM NaCl and 1 mM $MgCl_2$ in 50 mM Tris(Hydroxymethyl)aminomethane–HCl, pH 7.8 (1 g gelatine in 1 litre of TBS autoclaved for 30 min at 120 °C).

j Add TBS for B-values and an excess of hormone (we use 100 pmol) for ub-values.

k Incubate in the dark for IAA or ABA tests.

l Dilute alkaline phosphatase labelled tracer hormone from stock (kept in TBS/50% glycerol at −18 °C) with TBS-gelatine solution.

m Calculate as exemplified in Note 7, Protocol I. McAb adsorbed and alkaline phosphatase tracer dilution should be balanced so that the incubation with substrate results in an O.D. of ca. 1 when measured at 405 nm after an incubation with substrate for 1 h at 37 °C, see Fig. 11, for an example.

Acknowledgements

The author's own work was funded by grants of the Deutsche Forschungsgemeinschaft, Bonn, FRG, the Fonds der Chemischen Industrie, Frankfurt, FRG, the Stiftung Volkswagenwerk, Wolfsburg, FRG and the Stifterverband für die Deutsche Wissenschaft, Essen, FRG. We wish to thank all colleagues who have contributed by providing rare hormones and hormone analogues and we express our gratitude to Dr John Davies, Reading, U.K. for checking the English.

KEY REFERENCES

Goding, J. W. (1983). *Monoclonal Antibodies: Principles and Practice*. London: Academic Press.

Landsteiner, K. (1945). *The Specificity of Serological Reactions*. New York: Dover Publications.

Reeve, D. R. & Crozier, A. (1980). Quantitative analysis of plant hormones. In *Hormonal Regulation of Development I*, Encyclopedia of Plant Physiology, New Series, vol. **9**, ed. J. MacMillan, pp. 203–80. Berlin: Springer.

Yokota, T., Murofushi, N. & Takahashi, N. (1980). Extraction, Purification and Identification. In *Hormonal Regulation of Development I*, Encyclopedia of Plant Physiology, New Series, vol. **9**, ed. J. MacMillan, pp. 113–201. Berlin: Springer.

REFERENCES

Akiyoshi, D. E., Morris, R. O., Hinz, R., Mischke, B.S., Kosuge, T., Garfinkel, D. J., Gordon, M. P. & Nester, E. W. (1983). Cytokinin/auxin balance in crown gall tumors is regulated by specific loci in the T-DNA. *Proceedings of the National Academy of Sciences, U.S.A.*, **80**, 407–11.

Atzorn, R. & Weiler, E. W. (1983a). The immunoassay of gibberellins. 1. Radioimmunoassays for the gibberellins A_1, A_3, A_4, A_7, A_9 and A_{20}. *Planta*, **159**, 1–6.

Atzorn, R. & Weiler, E. W. (1983*b*). The immunoassay of gibberellins. 2. Quantitation of GA_3, GA_4 and GA_7 by ultrasensitive solid-phase enzyme immunoassays. *Planta*, **159**, 7–11.

Badenoch-Jones, J., Letham, D. S., Parker, C. W. & Rolfe, B. G. (1984). Quantitation of cytokinins in biological samples using antibodies against zeatin riboside. *Plant Physiology*, **75**, 1117–25.

Berson, S. A. & Yalow, R. (1959). Recent studies on insulin-binding antibodies. *Annals of the New York Academy of Sciences*, **82**, 338–44.

Cleland, R. E. (1983). Is plant development regulated by changes in the concentration of growth substances or by changes in the sensitivity to growth substances? *Trends in Biochemical Sciences*, October, 354–7.

Cutting, J. G., Lishman, A. W., van der Hoven, A. & Wolstenholme, B. N. (1983). Development of a radioimmunoassay for the cytokinin, isopentenyladenosine. *Crop Production*, **12**, 133–5.

Daie, J. & Wyse, R. (1982). Adaptation of the enzyme-linked immunosorbent assay (ELISA) to the quantitative analysis of abscisic acid. *Analytical Biochemistry*, **119**, 365–71.

deGreef, W., Dekegel, M. & Hamers, R. (1980). A radioimmunoassay for cytokinins. *Archives Internationales de Physiologie et Biochimie*, **88**, B134.

Eberle, J., Arnsheidt, A., Klixt, D. & Weiler, E. W. (1986). Monoclonal antibodies to plant growth regulators III. Zeatinriboside and dihydrozeatinriboside. *Plant Physiology*, in press.

Erlanger, B. F. (1980). The preparation of antigenic hapten-carrier conjugates: a survey. In *Methods of Enzymology*, vol. **70**, eds. H. van Vunakis & J. J. Langone, pp. 85–104. New York: Academic Press.

Ernst, D., Schäfer, W. & Oesterhelt, D. (1983). Isolation and quantitation of isopentenyladenosine in an anise cell culture by single-ion monitoring, radioimmunoassay and bioassay. *Planta*, **159**, 216–21.

Fuchs, S., Haimovich, J. & Fuchs, Y. (1971). Immunological studies of plant hormones. Detection and estimation by immunological assay. *European Journal of Biochemistry*, **18**, 384–90.

Fuchs, S. & Fuchs, Y. (1969). Immunological assay for plant hormones using specific antibodies to indoleacetic acid and gibberellic acid. *Biochimica et Biophysica Acta*, **192**, 528–30.

Fuchs, Y., Mayak, S. & Fuchs, S. (1972). Detection and quantitative determination of abscisic acid by immunological assay. *Planta*, **103**, 117–25.

Hansen, C. E., Wenzler, H. & Meins, F. Jr. (1984). Concentration gradients of trans-zeatinriboside and trans-zeatin in the maize stem. *Plant Physiology*, **75**, 959–63.

Hurn, B. A. L. & Chantler, S. M. (1980). Production of reagent antibodies. In *Methods of Enzymology*, vol. **70**, ed. H. van Vunakis & J. J. Langone, pp. 104–41. New York: Academic Press.

Kannangara, T., Simpson, G. M., Rajkumar, K. & Murphy, B. D. (1984). Analysis of abscisic acid in wheat leaves by a combination of high-performance liquid chromatography and radioimmunoassay. *Journal of Chromatography*, **283**, 425–30.

Khan, S. A., Humajun, M. Z. & Jacob, T. M. (1977). A sensitive radioimmunoassay for isopentenyladenosine. *Analytical Biochemistry*, **83**, 632–5.

Knöfel, H.-D., Bruckner, C., Sembdner, G. & Schreiber, K. (1984). A radioimmunoassay for jasmonic acid. *Biochemie und Physiologie der Pflanzen*, **179**, 317–25.

Köhler, G. & Milstein, C. (1975). Continuous culture of fused cells secreting antibody of predefined specificity. *Nature*, **256**, 495–7.

Le Page-Degivry, M. T., Duval, D., Bulard, C. & Delaage, M. (1984). A radioimmunoassay for abscisic acid. *Journal of Immunological Methods*, **67**, 119–28.

Leroux, B. (1984). *Recherche methodologique sur le dosage immunoenzymatique de trois substances de croissance vegetales, l'auxine, l'acide abscissique et la zeatine-riboside*. Thesis, Paris 1984.

MacDonald, E. M. S., Akiyoshi, D. E. & Morris, R. O. (1981). Combined high-performance liquid chromatography-radioimmunoassay for cytokinins. *Journal of Chromatography*, **214**, 101–9.

MacDonald, E. M. S. & Morris, R. O. (1983). Rapid, single-pass isolation of analytically pure plant hormones by immunoaffinity chromatography. *Plant Physiology* Supplement **72**, 26.

Mertens, R., Deus-Neumann, B. & Weiler, E. W. (1983). Monoclonal antibodies for the detection and quantitation of the endogenous plant growth regulator, abscisic acid. *FEBS Letters*, **160**, 269–72.

Mertens, R., Eberle, J., Arnscheidt, A., Ledebur, A. & Weiler, E. W. (1985). Monoclonal antibodies to plant growth regulators II. Indole-3-acetic acid. *Planta*, **166**, 389–93.

Mertens, R., Stüning, M. & Weiler, E. W. (1982). Metabolism of tritiated enantiomers of abscisic acid prepared by immunoaffinity chromatography. *Naturwissenschaften*, **69**, 595–6.

Mertens, R. & Weiler, E. W. (1983). Kinetic studies on the redistribution of endogenous growth regulators in gravireacting plant organs. *Planta*, **158**, 339–48.

Nemoto, N., Kawaoi, A., Okano, T., Ushiyama, H., Satoh, H. & Shikata, T. (1983). Immunohistochemistry of serotonin (5-HT) in the human gut endocrine cells. *Acta Histochemica et Cytochemica*, **16**, 577–87.

Pengelly, W. L., Bandurski, R. S. & Schulze, A. (1981). Validation of a radioimmunoassay for indole-3-acetic acid using gas-chromatography-selected ion monitoring-mass spectrometry. *Plant Physiology*, **68**, 96–8.

Pengelly, W. L. & Meins, F. Jr. (1977). A specific radioimmunoassay for nanogram quantities of the auxin, indole-3-acetic acid. *Planta*, **136**, 173–80.

Regier, D. A. & Morris, R. O. (1982). Secretion of trans-zeatin by *Agrobacterium tumefaciens*, a function determined by the nopaline Ti-plasmid. *Biochemical and Biophysical Research Communications*, **104**, 1560–6.

Rubenstein, K. E., Schneider, R. S. & Ullman, E. F. (1972) "Homogenous" enzyme immunoassay. A new immunochemical technique. *Biochemical and Biophysical Research Communications*, **47**, 846–51.

Schröder, G., Waffenschmidt, S., Weiler, E. W. & Schröder, J. (1984). The T-region of Ti-plasmids codes for an enzyme synthesizing indole-3-acetic acid. *European Journal of Biochemistry*, **138**, 387–91.

Sotta, B., Sossountzov, L., Maldiney, R., Sabbagh, I., Tachon, P. & Miginiac, E. (1985). Abscisic acid localization by light microscopic immunohistochemistry in *Chenopodium polyspermum* L. *Journal of Histochemistry and Cytochemistry*, **33**, 201–8.

Southwick, S. M., Davenport, T. L. & Castro, A. L. (1984). Development of radioimmunoassays for GA_3 and (+)ABA for use in the study of flower induction in "Tahiti" lime (*Citrus latifolia Tan.*). In *Proceedings of the 11th Meeting of the Plant Growth Regulator Society of America*, p. 59. Boston: PGRSA.

Treimer, J. F. & Zenk, M. H. (1978). Enzymic synthesis of Corynanthe-type alkaloids in cell cultures of *Cartharanthus roseus*: quantitation by radioimmunoassay. *Phytochemistry*, **17**, 227–31.

Trewavas, A. J. (1983). Is plant development regulated by changes in the concentration of growth substances or by changes in the sensitivity to growth substances? *Trends in Biochemical Sciences*, October, 354–7.

Trione, E. J. & Morris, R. O. (1983). Monoclonal antibodies against cytokinins. *Plant Physiology* Supplement, **72**, 114.

Ullman, E. F., Schwarzenberg, M. & Rubenstein, K. E. (1976). Fluorescence excitation transfer immunoassay. A general method for the determination of antigens. *Journal of Biological Chemistry*, **251**, 4172–8.

Vold, B. S. & Leonard, N. J. (1981). Production and characterization of antibodies and establishment of a radioimmunoassay for ribosylzeatin. *Plant Physiology*, **67**, 401–3.

Walton, D., Dashek, W. & Galson, E. (1979). A radioimmunoassay for abscisic acid. *Planta*, **146**, 139–45.

Wang, T. L., Futers, T. S., McGreary, F. & Cove, D. J. (1984). Moss mutants and the analysis of cytokinin metabolism. In *The biosynthesis and metabolism of plant hormones*, ed. A. Crozier & J.R. Hillman, pp. 135–64. Cambridge: Cambridge University Press.

Weiler, E. W. (1980*a*). Radioimmunoassay for the differential and direct analysis of free and conjugated abscisic acid in plant extracts. *Planta*, **148**, 262–72.

Weiler, E. W. (1980*b*). Radioimmunoassay for trans-zeatin and related cytokinins. *Planta*, **149**, 155–62.

Weiler, E. W. (1979). Radioimmunoassay for the determination of free and conjugated abscisic acid, *Planta*, **144**, 255–63.

Weiler, E. W. (1981). Radioimmunoassay for picomole quantities of indole-3-acetic acid for use with highly stable ^{125}I- and ^{3}H-IAA derivatives as radiotracers. *Planta*, **153**, 319–25.

Weiler, E. W. (1982). An enzyme-immunoassay for cis-(+)-abscisic acid. *Physiologia Plantarum*, **54**, 510–14.

Weiler, E. W., Jourdan, P. S. & Conrad, W. (1981). Levels of indole-3-acetic acid in intact and decapitated coleoptiles as determined by a specific and highly sensitive solid-phase enzyme immunoassay. *Planta*, **153**, 561–71.

Weiler, E. W., Schnabl, H. & Hornberg, C. (1982). Stress-related levels of abscisic acid in guard cell protoplasts of *Vicia faba* L. *Planta*, **154**, 24–28.

Weiler, E. W. & Spanier, K. (1981). Phytohormones in the development of crown gall tumors. *Planta*, **153**, 326–37.
Weiler, E. W. & Wieczorek, U. (1981). Determination of femtomole quantities of gibberellic acid by radioimmunoassay. *Planta*, **152**, 159–67.
Weiler, E. W. & Ziegler, H. (1981). Determination of phytohormones in phloem exudate of tree species by radioimmunoassay. *Planta*, **152**, 168–70.
Zaerr, J. B., Durley, R. C. & Morris, R.O. (1983). Applications of automated immunoaffinity chromatography to the analysis of cytokinins. *Plant Physiology* Supplement **72**, 26.
Zavala, M. E. & Brandon, D. L. (1983). Localization of a phytohormone using immunocytochemistry. *The Journal of Cell Biology*, **97**, 1235–9.

C. PERROT-RECHENMANN and P. GADAL

Enzyme immunocytochemistry

Introduction

Enzymes are proteins which may bear specific prosthetic groups, they also can undergo a great variety of post translational modifications such as glycosylation, phosphorylation and hydroxylation. In biochemical or physiological studies, enzyme amounts are generally measured in activity units rather than directly as proteins and the determination of the actual amount of the enzyme is almost never considered.

As organisms, higher plants are composed of organs – roots, stems, leaves and flowers. Each organ is made of different tissues, for instance, epidermis, stomata, mesophyll, phloem, xylem and collenchyma for the leaf. These tissues exhibit specific physiological properties, for example, photosynthesis in the mesophyll. Each tissue must then contain a specific set of enzymes different from those present in other tissues. At the cellular level, plant cells are rather complex and contain a number of different organelles such as the nucleus, chloroplasts, mitochrondria, vacuoles and cell walls in which are housed either specific enzymes or isoenzymes.

In order to understand properly the physiology of the plant, a knowledge of the precise localization of an enzyme at the tissue, cellular and subcellular levels is required. Up to now localization studies have been done mainly through enzyme determinations on isolated tissues or on separated organelles such as the nucleus, chloroplasts, mitochondria, vacuoles and many cases it cannot be applied to all plant tissues and, furthermore, the heterogeneity of the population of isolated organelles as well as their possible damage during isolation are still major problems. From the point of view of localization, enzyme cytochemistry represents an important improvement as it visualizes enzymes at their normal sites within cells. However, this method, elegant in its principle, presents severe limitations which are discussed in detail by Sexton & Hall (1978).

As proteins, enzymes are antigens. When such molecules are introduced into a vertebrate animal, its immune system will produce antibodies which react specifically with the antigen that induced their formation. As a consequence of their specificity, antibodies represent powerful probes to study plant proteins in general and plant enzymes more specifically.

It is important to stress the point that immunochemical detection or quantification on one hand and enzymatic activity on the other hand are not always correlated. Antibodies will detect the presence of a protein as such, even if for some reason, i.e. maturation or modification, the protein does not perform as an enzyme. On the other hand, activity means that the protein has been fully processed. Of course, enzyme activity implies the presence of the protein.

The major advantage of immunocytochemical techniques is that they are potentially suitable for all enzymes even if isoenzymes are present. They may relate also a protein-enzyme to a specific cell, cell type or organelle in a given tissue. Such a precise localization would not be possible using biochemical approaches.

General principles

Coons & Kaplan (1950) were the first to use a fluorescent antibody method in studies of animal cells. Later Singer (1959) and Singer & Schick (1961) introduced a metalloprotein (ferritin) as an antibody marker for electron microscopy. Compared to animal biologists, plant cell scientists have been slow to utilize these methods and it was not until 1970 that the first fluorescent antibody studies describing the localization of plant reserve proteins were reported (Graham & Gunning, 1970; Knox, Heslop-Harrison & Reed, 1970).

The reasons for such a long delay arose from the difficulties that they had to overcome before adapting the techniques to plant tissues. They include: isolation of pure enzyme, animal immunization, purification of the antibodies, preparation of plant sections suitable for antigen localization, choice of the detection system and extinction of the autofluorescence.

As a consequence of these difficulties, up to 1978 the localization of only four enzymes, namely α-amylase (Jacobsen & Knox, 1973), urease (Murray & Knox, 1977), cellulase (Bal *et al.*, 1976) and RuBP carboxylase (Hattersley, Watson & Osmond, 1977) had been investigated using these techniques.

Isolation of pure enzyme

No general method is available for the extraction and purification of plant enzymes used as antigenic materials (Rhodes, 1977). Usually it is necessary to establish the optimum conditions for each enzyme. However, general guidelines can now be formulated. First of all, care must be taken to homogenize plant material properly, especially for those tissues made of cells with thick cell walls such as guard cells or bundle sheath cells in C_4 plants. The composition of the extraction medium is a crucial factor. The pH for optimum activity of the enzyme is often used in view of the greater stability of many enzymes at that pH. In addition to a buffer, extraction media commonly include a thiol compound (ß-mercaptoethanol), EDTA, stabilizing ions and coenzymes.

Several low molecular weight substances, such as organic acids, tannins or phenolic compounds are released during grinding and may then interact covalently with proteins, the worst being polyphenols. Subsequently during immunization, they may behave as haptens and induce antibodies unrelated to the enzyme. The strategy to overcome that problem is to prevent oxidation and interactions between proteins and quinones during homogenization. This necessitates including effective protective agents such as polyvinylpyrrolidone or polyethylene glycol and minimizing conditions which stimulate oxidative reactions – extraction under nitrogen and inhibition of polyphenoloxidase by a copper chelating agent.

Although it is difficult to assess, the intervention of proteases is also probably very common and for some enzymes the presence of a cocktail of protease inhibitors is essential for a substantial recovery extraction (Wallace, 1977). Extraction time should be as short as possible and one should proceed very rapidly to the first step of enzyme purification in order to eliminate the maximum amount of harmful substances present in crude extracts.

As crude extracts of plant tissues are generally rather dilute in protein content, a concentration step is usually needed at an early stage in purification (Rhodes, 1977). Ammonium sulphate precipitation is widely used but a variety of other methods are available such as precipitation at the isoelectric point, lyophilization, gel exclusion, ultracentrifugation, ultrafiltration and polyethylene glycol precipitation. In choosing the concentration method, it should be kept in mind that the volumes are usually quite large.

The most important properties of proteins exploited by chromatographic methods are the molecule sizes and the ionic properties. Gel permeation chromatography is very efficient especially if the enzyme to be studied has a molecular weight different from the average values of plant enzymes (60–100 Kd); numerous matrixes are commercially available (e.g. Sephadex and Ultrogel; Reiland, 1971). Ion exchange chromatography has proved a powerful tool in protein purification since the introduction of the substituted cellulose by Peterson and Sober (1966). A range of substituted matrices – celluloses, dextrans and polyacrylamide – are available and have found wide applications (Himmelhoch, 1971). Other forms of chromatography include the use of alumina or various forms of calcium phosphate such as hydroxyapatite; these adsorbents can easily be used in batch forms (Scopes, 1978).

Another form of chromatography which is being more and more used with success is affinity chromatography or ligand-specific chromatography (Trayer, 1978). The principle is that the biological specificity for complementary molecules of an enzyme or a macromolecule can be utilized for a variety of applications when one of the components is immobilized onto an inert support. This technique offers tremendous possibilities for achieving separations which are difficult with less specific procedures. Columns are commercially available, but when specific ligands are needed, laboratory-made columns are not so difficult to prepare and are quite efficient. Immunoabsorption represents another form of affinity chromatography. Immunoadsorbents can be prepared by linking the purified immunoglobulins to an insoluble matrix, i.e. Sepharose or polyacrylamide, by a cyanogen bromide procedure or by a bifunctional reagent such as glutaraldehyde. Coupling through glutaraldehyde seems preferable to cyanogen bromide as some antibodies are released from the column on washing. As suggested by Mayer & Walker (1978), columns of adsorbents prepared from non-immune immunoglobulin may be used in series with the immunoadsorbent columns to remove non-specific binding species.

The kinetics for optimal binding of an antigen to an immunoadsorbent must be determined with respect to fixation time, to the pH and the ionic strength of the solution. If an enzyme is to be isolated from plant tissues with minimum denaturation, the immunoadsorbent should be prepared with an antiserum containing IgG with a low affinity to the antigen. Antisera do contain a mixture of IgGs, and their affinity, which is difficult to assess, varies from one preparation to another. Monoclonal antibodies (McAb) then seem more appropriate for this purpose.

Elution of the enzyme can be performed by a variety of agents, the aim of these treatments being to break the electrostatic and hydrophobic interactions which bind antigens to antibodies; low pH, high pH, high salt concentrations and chaotropic agents have all been used (Fornstedt, 1984). These conditions are often denaturing resulting in an inhibition or dissociation of the enzyme into its subunits. For immunization purposes, the dissociation does not impede antibody production as subunits are suitable for animal immunization. However, in some instances, the active enzyme can be eluted from the column using high pH (Paus, 1976), high salt concentration (Paus, 1976), or pure water (Vidal, Godbillon & Gadal, 1980).

The use of high-pressure liquid chromatography (HPLC) to purify almost any type of molecule has revolutionized separation technology in the past ten years (Perrett, 1978; Regnier, 1983). In this system, mobile phase is forced through a bed of microparticulate packing material at high velocity. The main advantage of HPLC is to obtain a high resolution in a minimum of time. However, HPLC is a more exacting technique than regular chromatography in that it requires great care in solvent purity, and operation of columns. Lack of attention to these details will yield less than optimum results. As antibody production requires milligram amounts of the antigen, the enzyme can be purified either by running several chromatographic separations or using a preparative column; on some of these columns up to 10 mg of protein can be loaded.

Electrophoresis especially polyacrylamide gel electrophoresis is one of the most widely used techniques in protein and enzyme chemistry (Gabriel, 1971*b*). Its main applications are in analysis of purity, detection of isozymes, genetic polymorphism and subunit molecular weight determinations. Moreover, electrophoresis and the related methods of isoelectric focusing and isotachophoresis can also be used with considerable success for enzymes (Wrigley, 1971). One of the major problems is to elute the protein from the gel. However, for animal injections, elution is not an absolute prerequisite as it is possible to cut out the band containing the enzyme and to inject the animal with the homogenized gel. When a protein solution is needed the enzyme can be electroeluted and collected on a dialysis membrane. Gel electrophoresis is a very powerful means of enzyme purification and could represent the final step in the purification of an enzyme.

Having passed the enzyme through a number of steps and brought it to a high state of homogeneity, it is then necessary to assess the degree of purity of the preparation. It should be realised that there is no formula which guarantees that a protein is pure. Over the years a number of tests of homogeneity have been applied; the best known are protein solubility in increasing salt concentrations, gel chromatography, ultracentrifugation and crystallisation. However, these methods usually require large amounts of protein (Phelps, 1978). As a consequence they have been largely replaced by high resolution analytical electrophoretic separation on polyacrylamide gel, by high resolution isoelectric focusing or better still, by two dimensional electrophoresis. Whenever possible, separations are used in conjunction with specific stains for proteins such as coomassie blue and specific tests to characterize enzyme activity (Gabriel, 1971*a*).

Another means which can be used to assess the purity of a protein preparation is by exploiting its immunological activity. Antibodies are raised to crude extracts of protein and are loaded opposite the enzyme at each purification step, resulting in immunodiffusion or immunoelectrophoresis and the production of a number of protein bands. As purification proceeds, the number of these bands should reduce until only a single precipitin line is viewed (Phelps, 1978).

Another criterion is the theoretical value of the specific activities. Further evidence can be provided from the number and the identity of N and C terminal amino acids in relation to the number of polypeptide chains in the protein molecule, if these are known. In this case, it should be remembered that proteases are very active in plant tissues and can chop off N terminal amino acids. At present, polyacrylamide gel electrophoresis within a pH range together with electrofocusing are the accepted standards to assess protein homogeneity.

Antiserum production

Rabbits are the most frequently used animal for antibody production. However, goats, sheep, rats, or mice are also suitable. The choice of the animal is of importance as the properties of the Fc fragment vary from species to species and, in immunocytochemical methods, almost all detection systems rely on the Fc fragment. Most plant proteins are not present in animals, as a consequence they exhibit a high antigenicity. As the immune response varies from animal to animal in a given species, it is better to inject more than one rabbit (usually three) in order to be able to select the best antiserum.

Antigen preparations may be presented to the animals by several procedures. In our protocol, 500 μl polyacrylamide-antigen-Freund's adjuvant mixtures are injected subcutaneously and intramuscularly on two occasions at two week intervals to rabbits "Fauve de Bourgogne" (2 × 300 μg antigen per rabbit) (Vidal *et al.*, 1980). Two intravenous booster injections of pure soluble antigen (2 × 500 μg) were given one month after the first injection at two day intervals. The rabbits were bled from the ear six days after the last injection. Blood is allowed to clot for 12 hours and the antiserum is carefully collected. Ammonium sulphate is added to the serum to give 33% saturation, the suspensions centrifuged and the IgG fractions further purified on a DEAE cellulose column equilibrated in 0.015 M phosphate buffer pH 7.8. The column is washed with the same buffer and the eluted IgG concentrated either by ammonium sulphate precipitation or diafiltration (Mayer & Walker, 1978). Another efficient way of purifying IgG is to make use of a protein A-Ultrogel column. After thoroughly washing the column, the IgG is eluted in citric acid/citrate buffer (pH 2.2 – 2.8) and concentrated. It must be noted that purified IgG is less stable than IgG present in the antiserum. It can be stabilized however, by adding a high concentration of albumin to the medium. Non-immune (control) serum should be similarly fractionated for use in comparative experiments with the IgG to the antigen of interest.

For antisera, it is essential to be sure of the specificity of the immune-serum in order to get reliable localization. Clearly the assessment of antiserum specificity is only as good as the sensitivity of the immunochemical analysis. It is then mandatory to use several techniques to assay for the specificity of the antibodies. Similarly the antiserum should be tested against the crudest tissue extract available as it precludes the possibility of missing some antigen reaction with antibodies in the antiserum.

Immunodiffusion and immunoelectrophoretic techniques are widely used to check antiserum specificity. They have already been excellently reviewed by Clausen (1981). Double diffusion analysis can be used to characterize the number of antigen-antibody systems with a tissue extract and a given antiserum. In classical immunoelectrophoresis, antigens are separated by electrophoresis in one dimension and then are allowed to interact by diffusion with the antiserum in a second dimension. This method provides a much more sensitive means testing antiserum specificity than immunodiffusion. Some enzymes retain their activity in immunoprecipitation lines (dehydrogenase, esterases, carboxylases). Thus, it is easy to identify the antigen-antibody complex of interest with the precipitin band.

An adaptation of the Southern blotting technique initially designed for DNA, is the attachment of fractionated protein to nitrocellulose paper (Towbin, Staehelin & Gordon, 1979; Burnette, 1981). The transfer from a polyacrylamide gel can be performed either by passive diffusion or electrophoresis. This method is referred to as "Western blotting" (see Thomas *et al.*, this volume). Of course, the monospecificity of the purified antibodies for a protein also must be assessed using the techniques described above. These purified antibodies are especially suitable for immunocytochemical localization as monospecificity is an absolute prerequisite for reliable localization of enzymes.

General strategies in optimizing antibody production

With the wide variety of possible procedures available, the beginner may well wonder how to proceed. When an enzyme has never or scarcely been purified it is highly recommended to extract small amounts of plant material and to pass the enzyme through different supports – gels, ion exchange, or hydroxyapatite – in order to learn more about the enzyme's physical properties, e.g. molecular weight and ionic properties. The methods yielding the highest increase in specific activity should be used in the final protocol. After these preliminary steps one should move to more specific methods such as affinity chromatography including immunoadsorption procedures. Polyacrylamide gel electrophoresis is probably the best way to get a pure enzyme and it is recommended that the band containing the protein of interest be cut out.

Antibodies can be prepared for commercially available enzyme preparations although antibodies raised against an enzyme extracted from the plant of interest are much better. However, very often antiserum multivalence will allow use of antibodies to a given enzyme for studies on the same enzyme in a related species. Whatever the origin of the antiserum, it is mandatory to assess very carefully its monospecificity if one does not want to get meaningless results. Very often extracts do contain different forms of the same enzyme. To localize the isoenzyme without ambiguity it is important to check for cross-reactivity and to be sure that the antibodies recognize only one of the isoforms. If some cross-reactivity is observed it is possible to remove the undesirable antibodies by adsorption of the antiserum with purified contaminating antigen(s) and to reassess the antiserum specificity (Fig. 1).

Fig. 1. General strategies for optimizing antibody production.

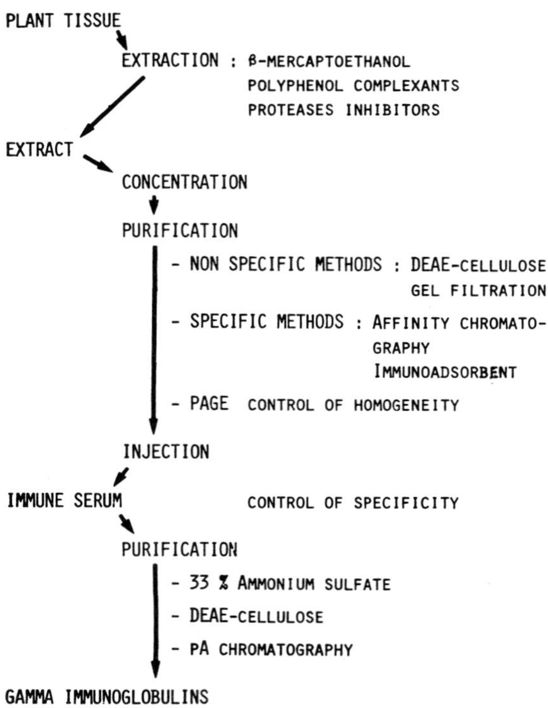

Applications

Green leaves of C_4 plants such as sorghum and maize exhibit a Kranz anatomy in which mesophyll and bundle sheath cells are structurally and functionally differentiated. In green leaf mesophyll, PEP carboxylase is involved in the primary CO_2 fixation resulting in oxaloacetate production which in turn is reduced to malate. The C_4 acid is transported and then decarboxylated in the bundle sheath, the resulting CO_2 being recaptured by RuBP carboxylase.

In maize leaves, cell division is restricted to the meristem located in the basal part of the leaf. Consequently, a gradient of differentiation exists from the base to the leaf tip. Studies were carried out either on samples taken all along the same leaf or on samples from the middle of leaves at various greening stages. In this section, we report immunofluorescence studies aimed to localize PEP carboxylase and RuBP carboxylase in the course of leaf ontogeny in these two C_4 plants.

PEP carboxylase localization

By using the antiserum directed against PEP carboxylase from green sorghum leaves on the basal meristem of green leaves or on samples taken from the middle part of very young leaves still inside the coleoptiles, fluorescence is emitted from all tissues: epidermis, mesophyll, bundle sheath and even xylem and phloem (Fig. 2.2). At the subcellular level, both etioplasts and cytosol were found fluorescent. It must however be pointed out that the antibodies used in this set of experiments were rather concentrated indicating that the enzyme concentration is rather low as already demonstrated by biochemical investigations (Vidal, Godbillon & Gadal, 1983). The conclusion to be drawn from these studies is that, in young leaves, there are no major differences between cells as far as PEP carboxylase is concerned.

In the course of leaf ontogeny, the major trend is an increase in total PEP carboxylase activity (Vidal *et al.*, 1983). Immunofluorescence studies indicate that the enzyme is concentrated in the cytosol of mesophyll cells and fades away gradually from other leaf cell compartments except in guard cells. For example, after 48 h of greening, fluorescence was almost restricted to the cytosol of mesophyll cells alone (Fig. 2.3).

If sections of 15-day old leaves which have never been illuminated are examined (Fig. 2.4), surprisingly the distribution pattern for PEP carboxylase is similar to the one found in green leaves. That is to say, high fluorescence is evident in the cytosol of mesophyll cells (Fig. 2.5). This result suggests that PEP carboxylase accumulation in the cytosol is not strictly light dependent but is part of an ontogenetic process whatever are

Fig. 2.1–2.6. Localization of PEP carboxylase in sections of sorghum and maize
2.1 and 2.6: control sections of etiolated leaf (2.1) and green leaf (2.6) treated with pre-immune serum (1/1000 dilution)
2.2. to 2.5: tissues treated with antiserum

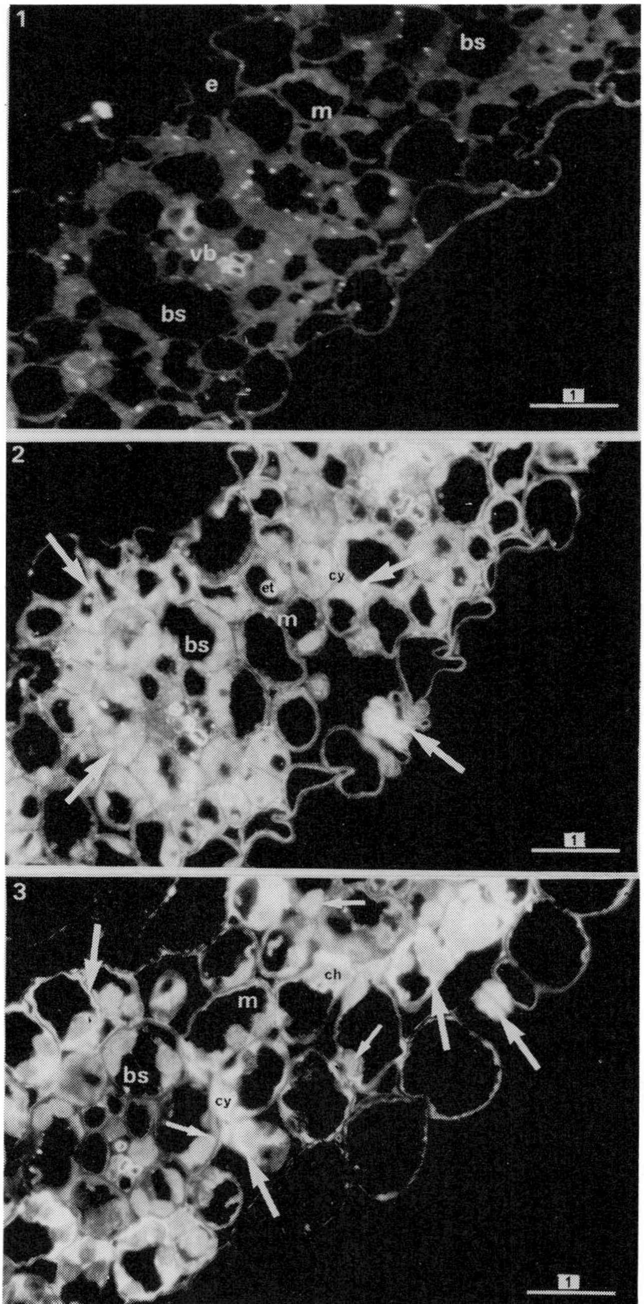

2.2. young etiolated leaf (1/1000 dilution of serum)
2.3. sorghum leaf, exposed for 48 hours to light (1/5000 dilution of serum)
2.4. etiolated leaf 15-days old (1/4000 dilution of serum)
2.5. green leaf 15-days old (1/10,000 dilution of serum)

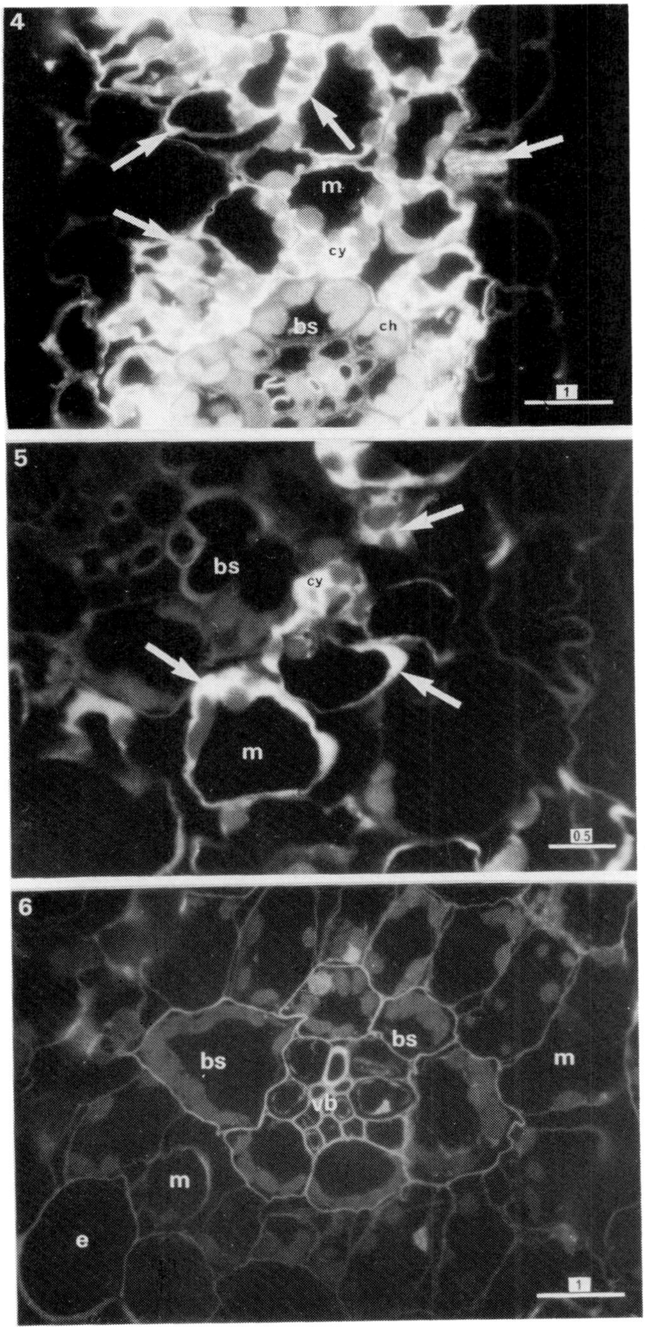

Fig. 2.7–2.9. Localization of RuBP carboxylase in sections of maize leaves. Tissues are treated with antiserum raised against a spinach enzyme
2.7. section of the basal meristem of a green leaf (1/500 dilution of the serum)
2.8. maize leaf exposed for 24 hours to light (1/1000 dilution of the serum)
2.9. green leaf (1/4000 dilution of the serum)

Abbreviations for Fig. 2: e = epidermal cells; m = mesophyll cells; bs = bundle sheath cells; vb = vascular bundle; ch = chloroplast; et = etioplast; cy = cytosol. Scale is given in μm.

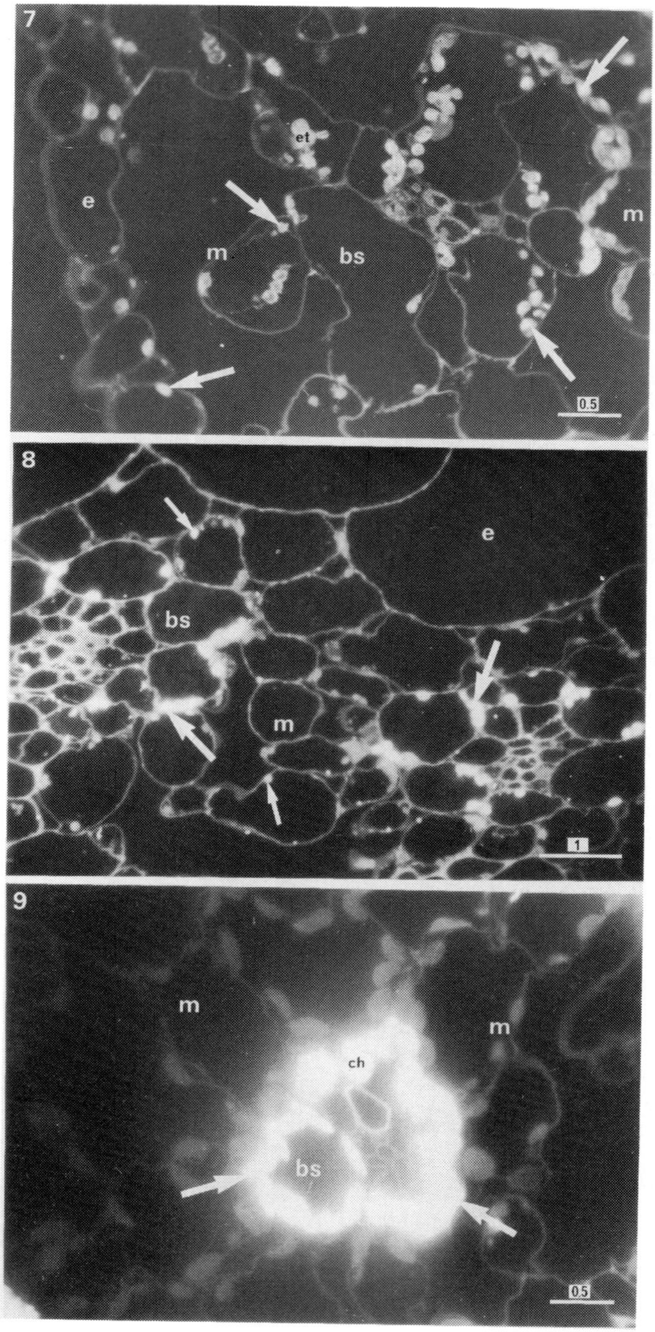

the external conditions. In this hypothesis, the light effect would just be to accelerate the differentiation events leading to a fully photosynthetically active leaf. Fig. 2.1 and Fig. 2.6 are control sections treated with pre-immune serum; no fluorescence can be detected.

RuBP carboxylase localization

In a second set of experiments, antiserum raised against RuBP carboxylase from spinach leaves was used to localize the enzyme in maize leaves. In leaves of three-day old seedlings or basal meristems, fluorescence was emitted from mesophyll and bundle sheath etioplasts (Fig. 2.7). However, the fact that the amount of antigen was very low is evident by the low dilution of the antibodies required to visualize the enzyme on the sections. Whether or not the enzyme is active in these very young leaves has not been determined in this study. Nevertheless, the results show the presence of the enzyme in dark grown leaves. In the course of greening, fluorescence disappears from mesophyll plastids and increases in bundle sheath cell etioplasts (Fig. 2.8). In completely differentiated green leaves, fluorescence is only displayed by the chloroplasts of the bundle sheath cells and the amount of antigen is much higher than in etiolated leaves (Fig. 2.9).

In the course of leaf differentiation and greening in C_4 plants, it is well established that an increase in PEP carboxylase and RuBP carboxylase occurs (Mayfield & Taylor, 1984). The immunolocalization techniques described in this paper show, in addition, that the increase in both carboxylases is restricted to specific tissues; mesophyll for PEP carboxylase, bundle sheath for RuBP carboxylase. At the subcellular level a compartmentation is also visualized, PEP carboxylase being in the cytosol and RuBP carboxylase in the chloroplasts. Another great advantage of the technique is that it allows the detection of small amounts of the enzyme in small samples. It represents, then, a complementary approach to the more classical biochemical techniques.

Preparation of plant sections suitable for antigen localization

For immunocytological studies, it is essential that the immunological properties of tissue proteins be preserved during the preparatory steps of the plant material. Specific immunoglobulins should have easy access and should be able to recognise and then bind to the antigen. Because of the plant cell wall, the post-embedding techniques are the most utilized (Fig. 3).

1. Fixation process

It is important that the chemical fixation alters the cellular constituents as little as possible. Aldehyde fixatives are often employed. Paraformaldehyde is preferentially used for immunofluorescence investigations. Glutaraldehyde ensures a good preservation of structures but it is autofluorescent and antigenic sites are less accessible. For other immunocytochemical localizations, a glutaraldehyde-paraformaldehyde mixture (Vreeland, 1972; Herman & Shannon, 1984*a*) or a glutaraldehyde-ethyldimethylaminopropyl carbodiimide mixture (Willingham & Yamada, 1979) give satisfactory results.

To prepare differentiated tissues with large vacuoles, as is the case for leaf parenchyma, it is recommended to check carefully the osmolarity of fixative and rinsing buffer; this avoids the plasmolysis of the cells. Plant tissues, especially leaves possess, on the one hand, epidermis covered by wax and gum and, on the other hand, thick pectocellulosic walls which hinder the good penetration of fixatives. To eliminate these problems, a partial vacuum is applied during fixation.

Fig. 3. Principal steps of post-embedding and pre-embedding procedures.

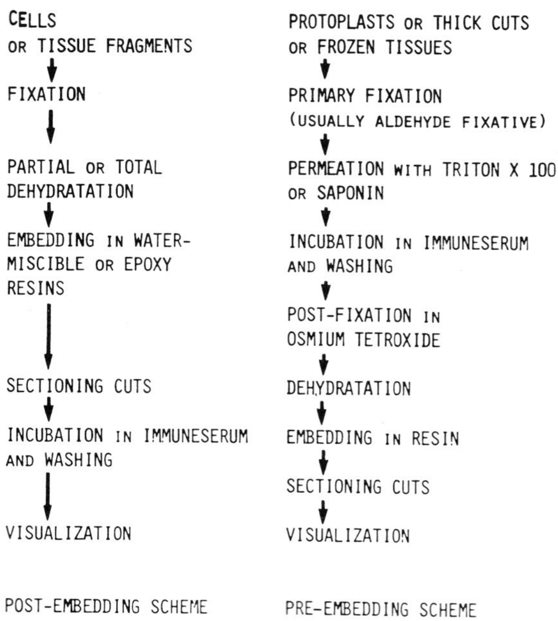

2. Dehydration

Dehydration has to be carried out very gradually in ethanol (or acetone) from 10% to 95% or absolute according to the type of embedding resin to be used. For tissues, such as crassulacean plant leaves, dehydration can be carried out in graded glycerol (Gadal, Perrot-Rechenmann & Vidal, 1983).

3. Embedding media

The choice of the embedding medium is also important. Although most resins have been tried with satisfactory results, it is not possible to generalize but rather each situation (i.e. type of antigen, its accessibility etc.) requires an appropriate medium. Two major types of resins should be distinguished:

(*a*) *Epoxy resins.* Embedding in epoxy resin requires total dehydration of the tissue. Polymerization is generally carried out at 60 °C. After sectioning (thin sections of 1 μm thick or ultrathin sections of 80–90 nm thick), the resin must be partially etched to facilitate access of antibody. The method of Mayor, Hampton & Rosario (1961) is especially efficient for the removal of araldite. Procedures suitable for epoxy resins are described by Herman & Shannon (1984*a*) for spurr, by Perrot *et al.* (1981) for epon/araldite mixture and by Perrot-Rechenmann *et al.* (1982) for araldite.

(b) Water miscible resins. In this case, tissues are partly dehydrated. Medium polymerization is carried out at low temperature and under UV light. The use of glycol methacrylate as an embedding medium for immunocytochemical investigations was described by Vreeland (1970) for brown seaweed then by Craig, Goodchild & Millerd, 1979 and Craig, Millerd & Goodchild, 1980 for pea seeds. However, this medium does not seem to be extensively used.

Lowicryl K4M is a relatively recent medium and presents several advantages. During dehydration and infiltration, samples can be kept in a partially hydrated state, since Lowicryl K4M may be polymerized with up to 5% water in the blocks (Carlemalm, Garavito & Villiger, 1982). Its use results in a better structural preservation (Roth *et al.*, 1981), improved preservation of antigenicity and significantly lower background labelling (Bendayan & Shore, 1982; Fruer, Wells & Ratcliffe, 1983; Whitehouse *et al.*, 1984; Herman & Shannon, 1984*b*, see Protocol section, Brewin *et al.*, this volume).

4. Cryotechniques

The cryotechniques, frequently employed to study animal tissues, can be applied to the preparation of plant material (Knox, Vithanage & Howlett, 1980; Jeffree, Yeoman & Kilpatrick, 1982). However, some problems inherent to the large vacuoles and to the thick cell wall of plant cells are difficult to overcome. The authors generally proceed by freeze substitution (Jones & Chen, 1976; Van Steveninck & Van Steveninck, 1978) or by freeze sectioning of the tissues (Roland, 1978; Jeffree & Yeoman, 1981; Burmeister & Hösel, 1981; Okamoto & Akazawa, 1979).

5. Pre-embedding procedure

In the case of thick frozen sections or of protoplasts for which cell walls are no longer an obstacle to the penetration of immunoglobulins, the pre-embedding techniques often used for animal tissues become suitable (Willingham & Yamada, 1979; Yamada, Yamada & Willingham, 1980; Franklin & Martin, 1981). The pre-embedding staining procedure is summarized in Fig. 3.

Choice of the immunocytochemical procedure

The following methods of investigation differ mainly on the visualization procedure, however, they can be described according to a general scheme.

1. Immunofluorescence

Immunofluorescence uses a fluorochrome as a marker, e.g. fluorescein isothiocyanate or tetramethylrhodamine isothiocyanate which are conjugated to an IgG. The major steps for an indirect procedure (Coons & Kaplan, 1950) are shown in Fig. 4. Sections were incubated with the native specific immune serum and then treated by fluorochrome stained antibodies raised against IgG previously fixed on the antigenic sites.

Direct immunofluorescence can be used too. The fluorochrome is then conjugated directly to specific antibodies. Some recognition can be masked and immune serum titre decreased. In addition, since the second antibody stage is not performed there is no amplification and, as a consequence, fluorescence detection is less sensitive.

The first studies on plant material based on immunofluorescence stressed a number of problems such as: tissue autofluorescence, fugacity of fluorochrome emission and non-specific reactions (Hattersley *et al.*, 1977; Craig *et al.*, 1979; Okamoto & Akazawa, 1979). The autofluorescence of plant tissues, studied by Nothnagle & Webb (1979) is often the main drawback. It is due to phenolic compounds, lignin, cellulose and chlorophylls. These compounds interfere more or less depending upon the choice of fluorochrome and on the wavelength selected for excitation. A treatment with a solution of Evans blue reduces the autofluorescence of the tissues sections. Under excitation at 495 nm, tissues coloured by Evans blue appeared red. In contrast, FITC emits yellow-green so the fluorescence due to specific labelling is easily detected and identified. The addition of paraphenylenediamine to the glycerol-buffer used for mounting the preparations has a marked effect on retarding the fluorescence fading during microscopic

Fig. 4. General steps for indirect immunofluorescence investigations.

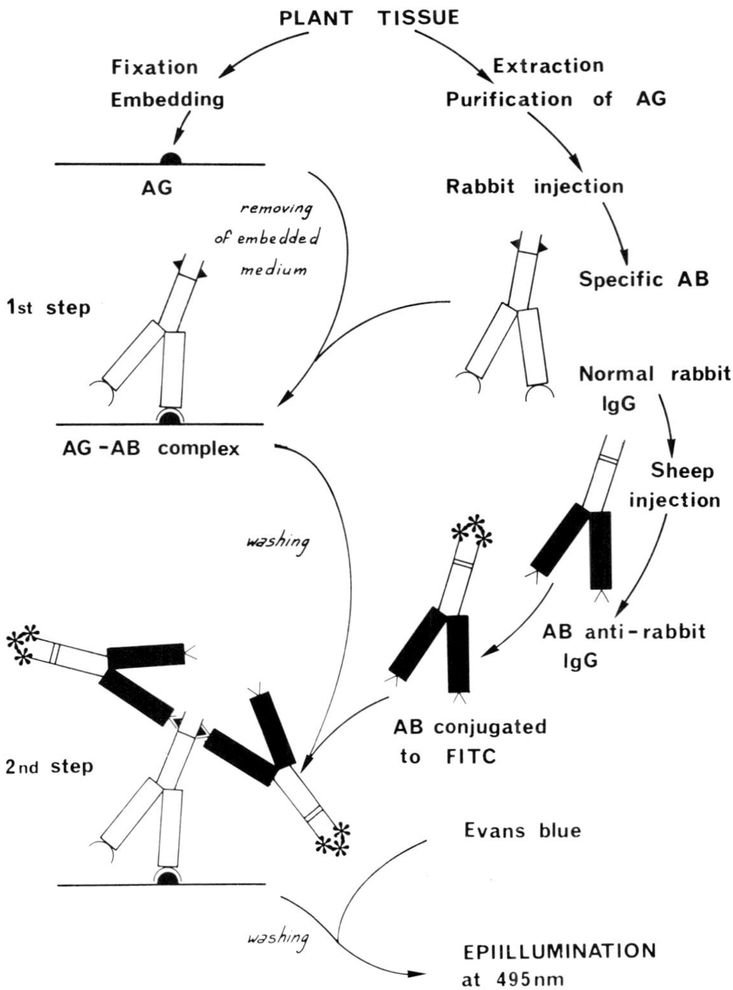

observation (Johnson & Aranjo, 1981). (For a detailed consideration of filter systems for fluorescence microscopy, see Roberts, this volume.)

Non-specific labelling can be extensively diminished by taking precautions such as:

(a) abundant and careful rinsing by phosphate buffer saline (PBS) or phosphate buffer saline tween (PBST).
(b) incubation of the sections with a decomplemented pre-immune serum (e.g. sheep serum, if the second antibodies against primary immunoglobulins were prepared in sheep). A serum is decomplemented by heating for 30 min at 56 °C. The aim of this treatment is to cover the specimens with proteins of the serum and in doing so to reduce non-specific adsorption.
(c) a gel filtration of the commercial fluorochrome labelled antibodies on a Sephadex G50 column eliminates significant amounts of contaminants.
(d) the use of a wet chamber for the incubations avoids specimen drying.

The preparations can be stored at most for 3–4 weeks at 4 °C in darkness.

Examples of the application of immunofluorescence methods to plant protein localization are: storage proteins in pea cotyledons (Craig *et al.*, 1979); lectin of *Datura stramonium* (Jeffree & Yeoman, 1981) and microtubule tubulins (Lloyd *et al.*, 1980; Wick *et al.*, 1981). Examples for enzymes – glutamine synthetase in spinach leaves and alnus root nodules (Hirel *et al.*, 1982*a*, *b*), NADP-malate dehydrogenase in corn leaves (Perrot-Rechenmann *et al.*, 1983), phosphoenolpyruvate carboxylase in leaves of higher plants (Perrot-Rechenmann *et al.*, 1982), RuBP carboxylase in leaves of C_3 and C_4 plants (Hattersley *et al.*, 1977), in leaves of *Panicum* species (Perrot-Rechenmann, Chollet & Gadal, 1984) and in guard cells (Madhavan & Smith, 1982) and urease in cotyledons of Jack beans (Murray & Knox, 1977).

2. *Immunoenzymatic methods*

Two methods can be used. The first one follows the same scheme as the indirect immunofluorescence approach. The enzyme is usually coupled to sheep immunoglobulin anti-rabbit IgG using bifunctional reagents such as glutaraldehyde (Nakane & Pierce, 1966; Avrameas, 1968). The most useful enzyme is peroxidase; glucose oxidase and alkaline phosphatase are less often employed. The second approach, the unlabelled antibody enzyme method, is characterized by using three levels of antibodies and peroxidase (Mason, Phifer & Spicer, 1969). The double use of antibody specificity in sequence depends on the bivalence of IgG. Peroxidase is specifically associated to its antibody as a soluble peroxidase-anti-peroxidase complex (Sternberger *et al.*, 1970). Fig. 5 shows the principal steps of these procedures. Peroxidase activity can be revealed in the presence of hydrogen peroxide and diaminobenzidine (Graham & Karnovsky, 1966) or hydrogen peroxide and 4-chloronaphthol (Nakane, 1968). These techniques are

Fig. 5. Unlabelled antibody peroxidase method.

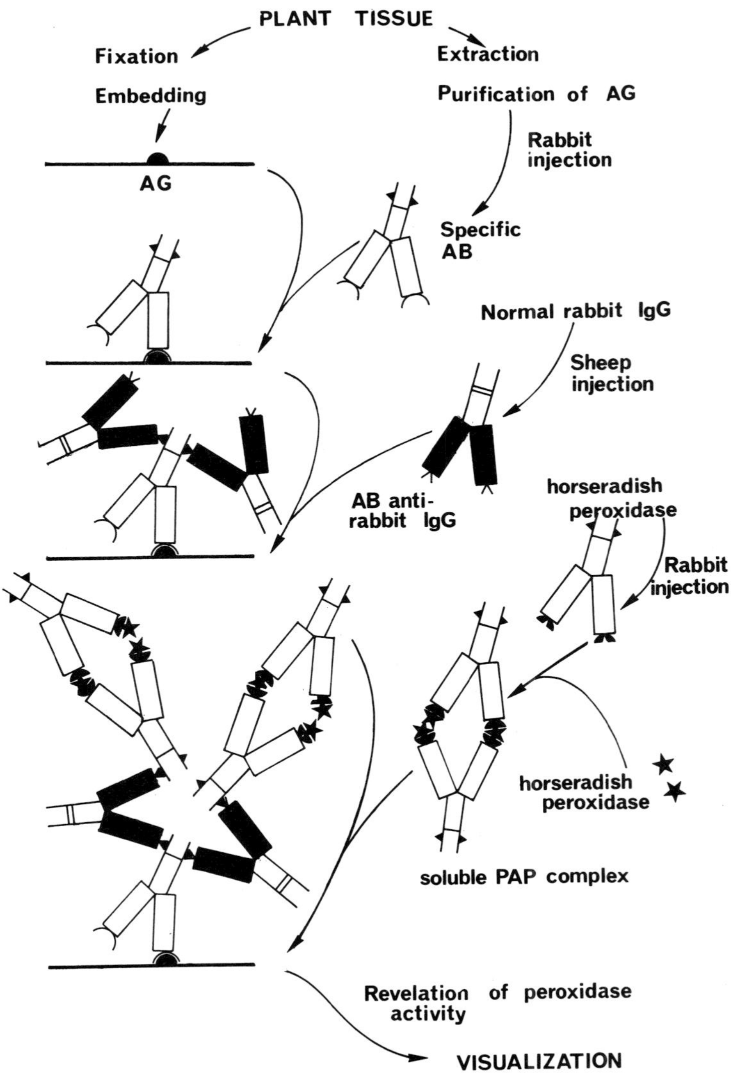

suitable for light and electron microscopy. The sensitivity of detection is high but the visualization of peroxidase activity can result in irregular precipitation. In some cases, especially for pre-embedding procedures, endogenous peroxidases prevent the use of these methods. Examples of the application of immunoperoxidase detection are: phytochrome in oat shoot (Pratt & Coleman, 1971), lectin in pea seeds (Van Driessche *et al.*, 1981), wheat germ agglutinin (Mishkind *et al.*, 1982), glycolipids in chloroplast membranes (Billecocq, 1976) and PEP carboxylase in sorghum leaves (Perrot *et al.*, 1981).

3. Indirect immunoferritin labelling

This method is adapted to electron microscopy investigations (Singer & Schick, 1961). A specific antiserum is used in native state and ferritin is linked with glutaraldehyde to antibodies against primary IgG (Otto, Takamiya & Vogt, 1973; Kismida *et al.*, 1975). Another procedure, in which specific antibodies are linked to biotinyl residues and visualized in electron microscopy by staining with avidin and biotinylated ferritin (Bayer, Wilcheck & Skutelsky, 1976; Heggeness & Ash, 1977) makes use of the very high binding affinity of biotin for avidin or streptavidin (the dissociation constant of avidin-biotin complex is 10^{10} M; Green, 1975). This results in an amplification of the primary reaction (Fig. 6). Nevertheless, there are few applications of these techniques. Ferritin particles are about 5–6 nm in diameter and their electron opacity is not very high. So, observations of preparations are difficult even after increasing the contrast by staining with uranyl acetate (Valentine, 1961) or lead citrate (Reynolds, 1963). Examples of this method applied to protein location are: storage protein in developing bean cotyledons (Baumgartner, Tokuyasu & Chrispeels, 1980), grass pollen allergen (Vithanage, Howlett & Knox, 1980), wheat germ agglutinin (Mishkind *et al.*, 1981), cellulase in pea stem (Bal *et al.*, 1976) and phosphoenolpyruvate carboxylase in sorghum leaves (Gadal *et al.*, 1983).

4. Immunogold staining

Colloidal gold used as an electron dense marker (Faulk & Taylor, 1971) presents numerous advantages. Gold granules have a high opacity to electrons and are good emitters of secondary electrons. These properties are required for markers in light (fluorescence), transmission (TEM) and scanning electron microscopy (SEM) (Geoghegan, Scilliam & Ackerman, 1978; Horisberger, 1978; Roth, Bendayan & Orci, 1978; De Mey, 1983). A SEM approach can be adapted to visualize cell surface receptors (Horisberger, Rosset & Bauer, 1975; Horisberger & Rosset, 1976, 1977; Mannweiler *et al.*, 1982).

In addition, monodisperse gold preparation can be obtained in a variety of sizes from tetrachloroauric acid by varying the reducing agent both qualitatively and quantitatively. The smallest granules (about 5 nm diameter) are produced with white phosphorus (Faulk & Taylor, 1971; Romano, Stokinski & Hughes-Jones, 1974) and granules of 12 nm average in diameter with sodium ascorbate (Stathis & Fabrikanos, 1958; Horisberger, Vonlanthen & Rosset, 1978). These two granules sizes are adapted to

Fig. 6. Immunoferritin labelling.

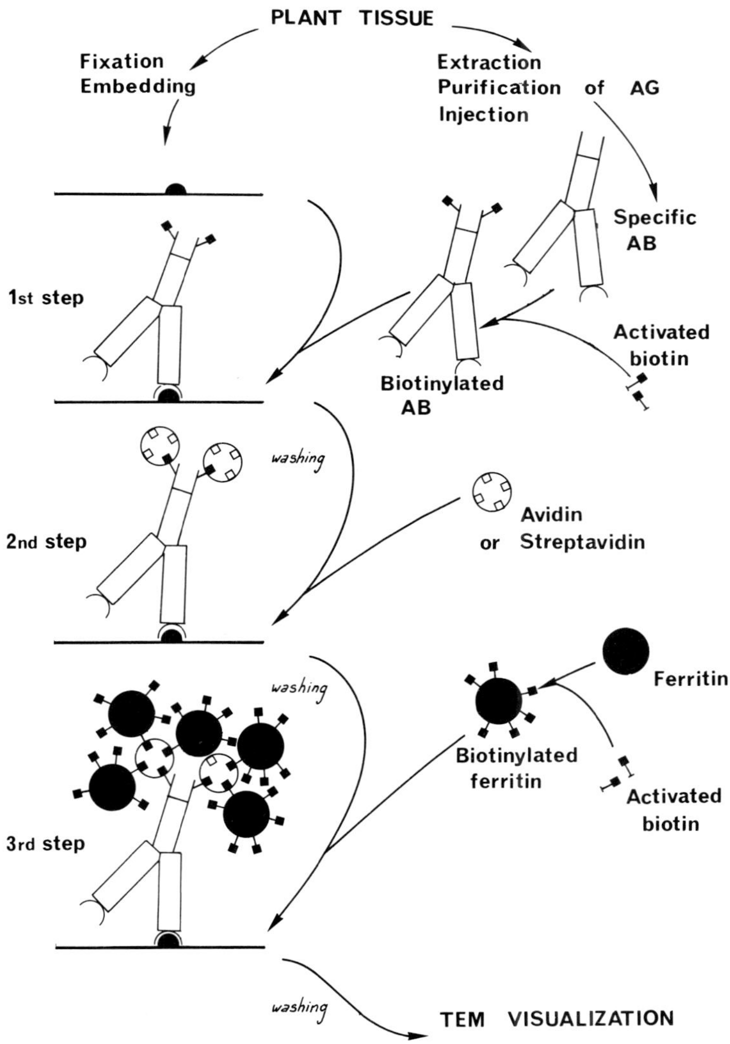

transmission electron microscopy investigation and are suitable for double labelling experiments (Horisberger & Vonlanthen, 1978; Geuze *et al.*, 1981; Bendayan, 1982). Large granules of 17 nm to 147 nm in diameter are prepared using decreasing amounts of sodium citrate and are usually employed for SEM (Frens, 1973; Horisberger & Rosset, 1977).

Colloidal gold can be coupled to IgG (of the second stage) (Romano *et al.*, 1974) or to protein A (pA) from *Staphylococcus aureus* (Romano & Romano, 1977; Roth, Bendayan & Orci, 1978). Protein A reacts with the Fc-region of IgG from several animal species; pA binds most subclasses of human, goat, rabbit, mouse and rat IgG but reaction is very low with sheep and cow IgG (Hjelm, Sjödahl & Sjöquist, 1975; Mota *et al.*, 1981; Findmark, Thoren-Tolling & Sjöquist, 1983). Examples of the applications of immunogold staining are – localization of galactomannan in *Schizosaccharomyces* (Horisberger *et al.*, 1978) and tubulin in *Haemanthus* endosperm (De Mey *et al.*, 1982).

All these investigations can be carried out with antisera or McAb for the primary step of antigen recognition. As we discussed previously, antisera raised against a given antigen are usually prepared in rabbits. On the other hand, McAb are obtained from mouse or rat. In this case, secondary antibodies are directed against rat or mouse immunoglobulin. The use of protein A is also possible. The IgG Fc-regions of these two species give a high reactivity with pA, except for rat IgG_{2a} which is not bound.

PROTOCOLS

I. Epoxy resin embedding and immunofluorescent analysis of plant leaves

1. Fix in 2% paraformaldehyde in 0.025 M mono disodium phosphate buffer, pH 7.3, for 3 h under a partial vacuum.
2. Wash for 2 h in 0.1 M phosphate buffer, pH 7.3, 0.4 M sucrose.
3. Dehydrate in 10% EtOH for 10 min, repeat twice;
4. Dehydrate in 25% EtOH for 10 min, repeat twice;
5. Dehydrate in 50% EtOH for 10 min, repeat twice;
6. Dehydrate in 75% EtOH for 10 min, repeat twice;
7. Dehydrate in 80% EtOH for 10 min, repeat twice;
8. Dehydrate in 90% EtOH for 4 × 10 min, repeat three times;
9. Dehydrate in absolute ethanol for 4 × 15 min, repeat three times.
10. Replace absolute ethanol by resin solvent, e.g. propylene oxide using 3 washes of 15 min.
11. Infiltrate with embedding resin, propylene-oxide:araldite; 3:1 for 1 h;
12. Infiltrate with resin, 1:1 for 1 h;
13. Infiltrate with resin, 1:3 overnight;

14. Infiltrate with resin, pure araldite for 12 h, twice.
15. Transfer tissue pieces and orientate in embedding moulds.
16. Polymerize at 37 °C in drying oven for 12 h, then at 60 °C for 3 days.
17. Mount thick sections of 1 μm in distilled water on glass slides.
18. Dry at 60 °C for a week.
19. Dip in sodium methoxide benzene for 3 min[a];
20. Dip in methanol-benzene mixture (1:1) for 1 min, twice;
21. Dip in acetone for 2 min, twice.
22. Dip in 95% EtOH for 5 min, twice;
23. Dip in 75% EtOH for 5 min, twice;
24. Rinse samples for 1 h under tap water.
25. Rinse 3 times 10 min each in 0.01 M phosphate buffer saline, pH 7.3[b].
26. Treat sections for 2 h at 20 °C with a decomplemented pre-immune sheep serum diluted 10 times[c].
27. Wash quickly in PBS and incubate for 60 h at 4 °C in drops of antiserum at various concentrations[d].
28. Wash the sections carefully three times for 10 min each in cold PBS.
29. Incubate for 45 min at 20 °C in FITC labelled sheep anti-rabbit IgG.
30. Rinse sections for 10 min, three times, in PBS and incubate for 15 min in 0.0001% (w/v) Evans blue diluted in PBS.
31. Rinse preparations thoroughly in PBS.
32. Mount in a glycerol-PBS mixture (8/2, v/v) containing 10 mg cm^{-3} of paraphenylenediamine.
33. Examine sections with an epiilluminating fluorescence microscope at 495 nm[e].

II. Preparation of gold stains

Preparation of Au_{12} granules

1. Add 10 cm^3 of 0.07% sodium ascorbate solution quickly at 20 °C to a rapidly stirred solution of 1 cm^3 of 1% $HAuCl_4$, 1 cm^3 of 0.2 M K_2CO_3 (to adjust pH to 6.9) and 25 cm^3 of distilled water.
2. Adjust the volume immediately to 100 cm^3 with distilled water when the mixture turns to a typical wine-red colour[f].

Preparation of pA-gold complex

1. Add the required amount of pA[g].
2. Homogenize the mixture for 5 min with a magnetic stirrer.
3. Add 1 cm^3 (for 20 cm^3) of 1% polyethylene glycol (Carbowax 20,000).
4. Centrifuge the suspension at 60,000 g for 50 min.
5. Resuspend the pelleted pA-gold complex in a small volume of PBS containing 1% polyethylene glycol 20,000.
6. Filter the preparation through 0.45 μm cellulose nitrate membranes and store at 4 °C.

Notes

a This and the following 5 steps processes the embedded material for immunofluorescence by removing the resin according to the method of Mayor *et al.* (1961).

b After removing the resin, sections should not dry otherwise non-specific reactions are observed.

c Serum is decomplemented by heating at 56 °C for 30 min.

d For control treatments, specific antiserum is replaced by pre-immune serum or PBS or an antiserum without specific antibodies (they can be removed by immunoprecipitation or immunoadsorption chromatography).

e Excitation at 495 nm is given by an ultra-high pressure mercury vapour lamp, and selective filters – BG 12 or interferential filters with narrow passing band (see Roberts, this volume, for details of filter systems).

f Colloidal gold granules are negatively charged and quite stable but increasing ionic strength by electrolyte addition generates colloid flocculation; this is visualized by the appearance of a blue colour in the solution.

g The minimal amount of pA for complete adsorption of colloidal gold is determined visually or spectrophotometrically by colour change from wine-red to blue as described in the Protocol (Roth & Binder, 1978). However, a 20% excess of pA is needed to stabilize colloidal gold suspension.

Acknowledgements

The authors are greatly indebted to Dr Jean Vidal for critically reading the manuscript and Dr Judy Brangeon for reviewing the English text. We thank Danielle Gillion and Françoise Maury for the typing.

KEY REFERENCES

Horisberger, M. & Rosset, J. (1977). Colloidal gold, as a useful marker for transmission and scanning electron microscopy. *The Journal of Histochemistry and Cytochemistry*, **25**, 295–305.

Jeffree, C. E., Yeoman, M. M. & Kilpatrick, D. C. (1982). Immunofluorescence studies on plant cells. *International Review of Cytology*, **80**, 231–65.

Rhodes, M. J. C. (1977). The extraction and purification of enzymes from plant tissues. In *Regulation of enzyme synthesis and activity in higher plants*, ed. H. Smith, pp. 245–69. London: Academic Press.

Scopes, R. K. (1978). Techniques for protein purification. In *Techniques in protein and enzyme biochemistry*, vol. B 1/I, part I, B 101, ed. H.L. Kornberg, J. C. Metcalfe, D. H. Northcote, C. I. Pogson & K. F. Tipton, pp. 1–42. Amsterdam, Oxford, New York: Elsevier.

Trayer, I. P. (1978). Affinity chromatography. Ibid, B 102, pp. 1–34.

REFERENCES

Avrameas, S. (1968). Detection d'anticorps et d'antigénes à l'aide d'enzymes. *Bulletin Société Chimie Biologique*, **50**, 1169–78.

Bal, A. K., Verma, D. P. S., Byrne, H. & Maclachlan, G. A. (1976). Subcellular localization of cellulases in auxin-treated pea. *The Journal of Cell Biology*, **69**, 97–105.

Baumgartner, B., Tokuyasu, K. T. & Chrispeels, M. J. (1980). Immunocytochemical localization of reserve protein in the endoplasmic reticulum of developing bean (*Phaseolus vulgaris*) cotyledons. *Planta*, **150**, 419–25.

Bayer, E. A., Wilcheck, M. & Skutelsky, E. (1976). Affinity cytochemistry: the localization of lectin and antibody receptors on erythrocytes via the avidin biotin complex. *FEBS Letters*, **68**, 240–4.

Bendayan, M. (1982). Double immunocytochemical labelling applying the protein A-gold technique. *The Journal of Histochemistry and Cytochemistry*, **30**, 1–81.

Bendayan, M. & Shore, G. G. (1982). Immunocytochemical localization of mitochondrial proteins in rat hepatocyte. *The Journal of Histochemistry and Cytochemistry*, **30**, 139–47.

Bernardi, G. (1971). Chromatography of proteins on hydroxyapatite. In *Methods in enzymology*, vol. **22**, ed. S. P. Colowick & N. O. Kaplan. pp. 325–39. New York & London: Academic Press.

Billecocq, A. (1976). Structures of membranes: localization of glycolipids in chloroplast membranes by immunoenzymatic techniques related to electron microscopy. *First International Symposium on Immunoenzymatic Techniques*, ed. G. Feldmann, P. Druet, J. Bignon & S. Avrameas. pp. 487–90. Amsterdam: Elsevier/North Holland.

Burmeister, G. & Hösel, W. (1981). Immunohistochemical localization of ß-glucosidases in lignin and isoflavone metabolism in *Cicer arietinum* L. seedlings. *Planta*, **152**, 578–86.

Burnette, W. N. (1981). "Western blotting" electrophoretic transfer of proteins from sodium dodecyl sulfate-polyacrylamide gels to unmodified nitrocellulose and radiographic detection with antibody and radio-iodinated protein A. *Analytical Biochemistry*, **112**, 195–203.

Carlemalm, E., Garavito, R. M. & Villiger, W. (1982). Resin development for electron microscopy and an analysis of embedding at low temperature. *Journal of Microscopy*, **126**, 123–43.

Clausen, J. (1981). *Immunochemical techniques for the identification and estimation of macromolecules*. 2nd ed., ed. T.S. Work & E. Work. Amsterdam, Oxford, New York: Elsevier/North Holland.

Coons, A. H. & Kaplan, M. H. (1950). Localization of antigen in tissue cells. Improvements in method for detection of antigen by means of fluorescent antibody. *Journal of Experimental Medicine*, **91**, 1–13.

Craig, S., Goodchild, D. J. & Millerd. A. (1979). Immunofluorescent localization of pea storage proteins in glycol methacrylate embedded tissue. *The Journal of Histochemistry and Cytochemistry*, **27**, 1312–16.

Craig, S., Millerd, A. & Goodchild, D. J. (1980). Structural aspects of protein accumulation in developing pea cotyledons. III. Immunocytochemical localization of legumin and vicilin antibodies shown to be specific by the enzyme linked immunosorbent assay. *Australian Journal of Plant Physiology*, **7**, 339–51.

De Mey, J. (1983). The preparation of immunoglobulin gold conjugates (IGS reagents) and their use as markers for light and electron microscopic immunocytochemistry. *IBRO Handbook Series*, **3**, 347–72.

De Mey, J., Lambert, A. M., Bajer, A. S., Moeremans, M. & De Brabander, M. (1982). Visualization of microtubules in interphase and mitotic plant cells in *Haemanthus* endosperm with the immunogold staining method. *Proceedings of the National Academy of Sciences, U.S.A.*, **79**, 1898–902.

Fornstedt, N. (1984). Affinity chromatographic studies on antigen-antibody dissociation. *FEBS Letters*, **177**, 195–99.

Faulk, W. P. & Taylor, G. M. (1971). An immunocolloid method for the electron microscope. *Immunocytochemistry*, **8**, 1081–3.

Findmark, R., Thoren-Tolling, K. & Sjöquist, J. (1983). Binding of immunoglobulins to protein A and immunoglobulin levels in mammalian sera. *Journal of Immunological Methods*, **61**, 1–13.

Franklin, R. M. & Martin, M. T. (1981). Pre-embedding immunohistochemistry as an approach to high resolution antigen localization at the light microscopic level. *Histochemistry*, **72**, 173–90.

Frens, G. (1973). Controlled nucleation for the regulation of the particle size in monodisperse gold solutions. *Nature Physical Science*, **241**, 20–2.

Fruer, P. R., Wells, C. & Ratcliffe, A. T. (1983). Technical difficulties overcome in the use of Lowicryl 4KM electron microscopic embedding resin. *Histochemistry*, **77**, 141–3.

Gabriel, O. (1971*a*). Locating enzymes on gels. In *Methods in Enzymology*, vol. **22**, ed. S. P. Colowick & N. O. Kaplan. pp. 578–605. New York, London: Academic Press.

Gabriel, O. (1971*b*). Analytical disc gel electrophoresis. In *Methods in Enzymology*, vol. **22**, ed. S. P. Colowick & N. O. Kaplan, pp. 565–78. New York, London: Academic Press.

Gadal, P., Perrot-Rechenmann, C. & Vidal, J. (1983). Immunocytochemical visualization of phosphoenol pyruvate carboxylase in higher plants. *Physiologie Végétale*, **21**, 1055–62.

Geoghegan, W. D., Scilliam, J. J. & Ackerman, G. A. (1978). The detection of human B lymphocytes by both light- and electron microscopy utilizing colloidal gold labelled anti-immunoglobulin. *Immunological Communications*, **7**, 1–12.

Geuze, H. J., Slot, J. W., Vanderley, P. A. & Scheffer, R. C. T. (1981). Use of colloidal gold particles in double-labelling immunoelectron microscopy of ultra-thin frozen tissue sections. *The Journal of Cell Biology*, **89**, 653–65.

Graham, R. C. & Karnovsky, H. J. (1966). The early stages of absorption of injected horse radish peroxidase in the proximal tubules of mouse kidney. Ultrastructural cytochemistry by a new technique. *The Journal of Histochemistry and Cytochemistry*, **14**, 291–302.

Graham, T. A. & Gunning, B. E. S. (1970). Localization of legumin and vicilin in bean cotyledon cells using fluorescent antibodies. *Nature*, **228**, 81–2.

Green, N. M. (1975). Avidin. *Advances in Protein Chemistry*, **29**, 85–133.

Hattersley, P. W., Watson, L. & Osmond, C. B. (1977). *In situ* immunofluorescent labelling of ribulose-1,5-bisphosphate carboxylase in leaves of C_3 and C_4 plants. *Australian Journal of Plant Physiology*, **4**, 523–39.

Heggeness, M. H. & Ash, J. F. (1977). Use of the avidin-biotin complex for the localization of actin and myosin with fluorescence microscopy. *The Journal of Cell Biology*, **73**, 783–8.

Herman, E. M. & Shannon, L. M. (1984*a*). Immunocytochemical localization of concanavalin A in developing jack-bean cotyledons. *Planta*, **161**, 97–104.

Herman, E. M. & Shannon, L. M. (1984*b*). Immunocytochemical evidence for the involvement of Golgi apparatus in the deposition of seed lectin of *Bauhinia purpurea* leguminosae. *Protoplasma*, **121**, 163–70.

Himmelhoch, S. R. (1971). Chromatography of proteins on ion-exchange adsorbents. In *Methods in Enzymology*, vol. **22**, ed. S. P. Colowick & N. O. Kaplan pp. 273–286. New York, London: Academic Press.

Hirel, B., Perrot-Rechenmann, C., Suzuki, A., Vidal, J. & Gadal, P. (1982*a*). Glutamine synthetase in spinach leaves. Immunological studies and immunocytochemical localization. *Plant Physiology*, **69**, 983–7.

Hirel, B., Perrot-Rechenmann, C., Maudinas, B. & Gadal, P. (1982*b*). Glutamine synthetase in Alder (*Alnus glutinosa*) root nodules. Purification, properties and cytoimmunochemical localization. *Plant Physiology*, **55**, 197–203.

Hjelm, J., Sjödahl, J. & Sjöquist, J. (1975). Immunologically active and structural similar fragments of protein A from *Staphylococcus aureus*. *European Journal of Biochemistry*, **57**, 395–403.

Horisberger, M. (1979). Evaluation of colloidal gold as a cytochemical marker for transmission and scanning electron microscopy. *Biologie Cellulaire*, **36**, 253–8.

Horisberger, M. & Rosset, J. (1976). Localization of wheat germs agglutinin receptors on yeast cells by scanning electron microscopy. *Experientia*, **32**, 998–1000.

Horisberger, M. & Rosset, J. (1977). Colloidal gold, a useful marker for transmission and scanning electron microscopy. *The Journal of Histochemistry and Cytochemistry*, **25**, 295–305.

Horisberger, M., Rosset, J. & Bauer, H. (1975). Colloidal gold granules as markers for cell surface receptors in the scanning electron microscope. *Experientia*, **31**, 1147–8.

Horisberger, M. & Vonlanthen, M. (1978). Simultaneous localization of a hepatic binding protein specific for galactose and of galactose-containing receptors on rat hepatocytes. *The Journal of Histochemistry and Cytochemistry*, **26**, 960–6.

Horisberger, M., Vonlanthen, M. & Rosset, J. (1978). Localization of galactomannan and of wheat germ agglutinin receptor in *Schizosaccharomyces pombe*. *Archives of Microbiology*, **119**, 107–11.

Jacobsen, J. V. & Knox, R. B. (1973). Cytochemical localization and antigenicity of amylase in barley aleurone tissue. *Planta*, **112**, 213–24.

Jeffree, C. E. & Yeoman, M. M. (1981). A study of the intracellular and intercellular distribution of the *Datura stramonium* lectin using an immunofluorescent technique. *The New Phytologist*, **87**, 463–71.

Jeffree, C. E., Yeoman, M. M. & Kilpatrick, D. C. (1982). Immunofluorescence studies on plant cells. *International Review of Cytology*, **80**, 231–65.

Johnson, G. D. & de Nogueiro Aranjo, G. M. (1981). A simple method for reducing the fading of immunofluorescence during microscopy. *Journal of Immunological Methods*, **43**, 349–50.

Jones, R. L. & Chen, R. F. (1976). Immunohistochemical localization of α-amylase in barley aleurone cells. *Journal of Cell Science*, **20**, 183–98.

Kismida, Y., Olsen, B. R., Berg, R. A. & Prockop, D. J. (1975). Two improved methods for preparing ferritin-antibody conjugates for electron microscopy. *The Journal of Cell Biology*, **64**, 331–9.

Knox, R. A., Heslop-Harrison, J. & Reed, C. (1970). Localization of antigens associated with the pollen grain wall by immunofluorescence. *Nature*, **225**, 1066–8.

Knox, R. B., Vithanage, H. I. M. V. & Howlett, B. J. (1980). Botanical immunocytochemistry. A review with special reference to pollen antigens and allogens. *The Histochemical Journal*, **12**, 247–72.

Leech, R. M. (1977). Subcellular fractionation techniques in enzyme distribution studies. In *Regulation of enzyme synthesis and activity in higher plants*, ed. H. S. Smith, pp. 289–327. New York, London: Academic Press.

Lloyd, C. W., Slabas, A. R., Powell, A. J. & Louve, S. B. (1980). Microtubules, protoplasts and plant cell shape. An immunofluorescent study. *Planta*, **147**, 500–6.

Madhavan, S. & Smith, B.N. (1982). Localization of ribulose bisphosphate carboxylase in the guard cells by an indirect immunofluorescence technique. *Plant Physiology*, **69**, 273–7.

Mannweiler, K., Hohenberg, H., Bohn, W. & Rutter, G. (1982). Protein A gold particles as markers in replica immunocytochemistry: high resolution electron microscope investigations of plasma membrane surfaces. *Journal of Microscopy*, **126**, 145–9.

Mason, T. E., Phifer, R. F. & Spicer, S. S. (1969). An immunoglobulin enzyme bridge method for localizing tissue antigens. *The Journal of Histochemistry and Cytochemistry*, **9**, 909–10.

Mayer, R. J. & Walker, J. H. (1978). Techniques in enzyme and protein immunochemistry. In *Techniques in protein and enzyme biochemistry*, vol. B 1/II, part

II, B **119**, ed. H. L. Kornberg, J. C. Metcalfe, D. H. Northcote, C. I. Pogson, & K. F. Tipton, pp. 1–32. Amsterdam, Oxford, New York: Elsevier/North-Holland.

Mayfield, S. P. & Taylor, W. C. (1984). The appearance of photosynthetic protein in developing maize leaves. *Planta*, **161**, 481–6.

Mayor, H. D., Hampton, J. C. & Rosario, B. (1961). A simple method for removing the resin from epoxy-embedding tissue. *Journal of Biophysical and Biochemical Cytology*, **9**, 909–10.

Mishkind, M., Raikhel, N. V., Palevitz, B. A. & Keegstra, K. (1982). Immunocytochemical localization of wheat germ agglutinin in wheat. *The Journal of Cell Biology*, **92**, 753–64.

Mota, G., Moraru, I., Sjöquist, J. & Ghetie, V. (1981). Protein A as a molecular probe for the detection of antigen induced conformational change in FC region of rabbit antibody. *Molecular Immunology*, **18**, 373–8.

Murray, D. R. & Knox, R. B. (1977). Immunofluorescent localization of urease in the cotyledons of Jack bean, *Canavalia ensiformis*. *Journal of Cell Science*, **26**, 9–18.

Nakane, P. K. (1968). Simultaneous localizations of multiple tissue antigen using the peroxidase-labelled antibody method: a study on the pituitary gland of the rat. *Journal of Histochemistry and Cytochemistry*, **16**, 557–60.

Nakane, P. K. & Pierce, G. B. (1966). Enzyme-labelled antibodies: preparation and application for the localization of antigens. *The Journal of Histochemistry and Cytochemistry*, **14**, 929–31.

Nothnagle, E. A. & Webb, W. W. (1979). Barrier filter for fluorescence microscopy of strongly autofluorescent plant tissues. *The Journal of Histochemistry and Cytochemistry*, **27**, 1000–2.

Okamoto, K. & Akazawa, T. (1979). Enzymic mechanism of starch breakdown in germinating rice seeds. Immunohistochemical localization of β-amylase. *Plant Physiology*, **64**, 337–40.

Otto, H., Takamiya, H. & Vogt, A. (1973). A two-stage method for cross linking antibody globulin to ferritin by glutaraldehyde. *Journal of Immunological Methods*, **3**, 137–46.

Paus, E. (1976). Immunoadsorbent affinity purification of the two enzyme forms of mannosidase from *Phaseolus vulgaris*. *FEBS Letters*, **72**, 39–42.

Perrett, D. (1978). Application of high performance liquid chromatography (HPLC) to biochemical analysis. In *Techniques in metabolic research*, vol. B 2/II, part II, B 215, ed. H. L. Kornberg, J. C. Metcalfe, D. H. Northcote, C. I. Pogson & K. F. Tipton, pp. 1–22. Amsterdam, Oxford, New-York: Elsevier/ North-Holland.

Perrot, C., Vidal, J., Burlet, A. & Gadal, P. (1981). On the cellular localization of phosphoenolypyruvate carboxylase in *Sorghum* leaves. *Planta*, **151**, 226–31.

Perrot-Rechenmann, C., Vidal, J., Brulfert, J., Burlet, A. & Gadal, P. (1982). A comparative immunocytochemical localization study of phosphoenolpyruvate carboxylase in leaves of higher plants. *Planta*, **155**, 24–30.

Perrot-Rechenmann, C., Jacquot, J. P., Gadal, P., Weeden, N. F., Cseke, C. & Buchanan, B. B. (1983). Localization of NADP-malate dehydrogenase of Corn leaves by immunological methods. *Plant Science Letters*, **30**, 219–26.

Perrot-Rechenmann, C., Chollet, R. & Gadal, P. (1984). *In situ* immunofluorescent localization of phosphoenolpyruvate and ribulose-1,5-bisphosphate carboxylases in leaves of C_3, C_4 and C_3–C_4 intermediate *Panicum* species. *Planta*, **161**, 266–71.

Peterson, E. A. & Sober, H. A. (1966). Column chromatography of proteins substituted celluloses. In *Methods in Enzymology*, vol. **5**, ed. S. P. Colowick & N. O. Kaplan, pp. 3–27. New York & London: Academic Press.

Phelps, C. (1978). The purity of proteins. In *Techniques in protein and enzyme biochemistry*, vol. B 1/I, Part I, B **103**, ed. H. L. Kornberg, J. C. Metcalfe, D. H. Northcote, C. I. Pogson, & K. F. Tipton, pp. 1–17. Amsterdam, Oxford, New York: Elsevier/North-Holland.

Pratt, L. H. & Coleman, R. A. (1971). Immunocytochemical localization of phytochrome. *Proceedings of the National Academy of Sciences, U.S.A.*, **68**, 2431–5.

Regnier, F. E. (1983). High-performance liquid chromatography of proteins. In *Methods in Enzymology*, vol. **91**, ed. S. P. Colowick & N. O. Kaplan, pp. 137–90. New York & London: Academic Press.

Reiland, J. (1971). Gel filtration. In *Methods of Enzymology*, vol. **22**, ed. S. P. Colowick & N. O. Kaplan, pp. 287–321. New York & London: Academic Press.

Reynolds, E. S. (1963). The use of lead citrate at high pH as an electron opaque stain in electron microscopy. *The Journal of Cell Biology*, **17**, 208–12.

Rhodes, M. J. C. (1977). The extraction and purification of enzymes from plant tissues. In *Regulation of enzyme synthesis and activity in higher plants*, ed. H. Smith, pp. 245–69. London: Academic Press.

Roland, J. C. (1978). General preparation and staining of thin sections. In *Electron microscopy and cytochemistry of plant cells*, ed. J.H. Hall, pp. 1–62. Amsterdam, Oxford, New York: Elsevier/North-Holland.

Romano, E. L. & Romano, M. (1977). Staphylococcal protein-A bound to colloidal gold – useful reagent to label antigen-antibody sites in electron microscopy. *Immunocytochemistry*, **14**, 711–15.

Romano, E. L., Stokinski, C. & Hughes-Jones, N. C. (1974). An antiglobulin reagent labelled with colloidal gold for use in electron microscopy. *Immunocytochemistry*, **11**, 521–2.

Roth, J., Bendayan, M., Carlemalm, E., Villiger, W. & Garavito, R. M. (1981). Enhancement of structural preservation and immunocytochemical staining in low temperature embedded pancreatic tissue. *The Journal of Histochemistry and Cytochemistry*, **29**, 663–71.

Roth, J., Bendayan, M. & Orci, L. (1978). Ultrastructural localization of intracellular antigens by the use of protein-A Gold complex. *The Journal of Histochemistry and Cytochemistry*, **26**, 1074–81.

Roth, J., Bendayan, M. & Orci, L. (1980). FITC-Protein-A-Gold complex for light and electron microscopic immunocytochemistry. *The Journal of Histochemistry and Cytochemistry*, **28**, 55–7.

Roth, J. & Binder, M. (1978). Colloidal gold, ferritin and peroxidases for electron microscopic double labelling lectin technique. *Journal of Histochemistry and Cytochemistry*, **26**, 163–9.

Scopes, R. K. (1978). Techniques for protein purification. In *Techniques in protein and enzyme biochemistry*, vol. B 1/I, part I, B **101**, ed. H. L. Kornberg, J. C. Metcalfe, D. H. Northcote, C. I. Pogson & K. F. Tipton, pp. 1–42, Amsterdam, Oxford, New York: Elsevier/North-Holland.

Sexton, R. & Hall, J. L. (1978). Enzyme cytochemistry. In *Electron microscopy and cytochemistry of plant cells*, ed. J. L. Hall, pp. 63–147. Amsterdam, Oxford, New York: Elsevier/North-Holland.

Singer, S. J. (1959). Preparation of an electron dense antibody conjugate. *Nature*, **183**, 1523–4.

Singer, S. J. & Schick, A. F. (1961). The properties of specific stains for electron microscopy prepared by conjugation of antibody molecules with ferritin. *Journal of Biophysical and Biochemical Cytology*, **9**, 519–25.

Slot, J. W. & Geuze, H. J. (1981). Sizing of protein A-colloidal gold probes for immunoelectron microscopy. *Journal of Cell Biology*, **90**, 533–6.

Stathis, E. C. & Fabrikanos, A. (1958). Preparation of colloidal gold. *Chemistry and Industry*, **27**, 860–1.

Sternberger, L. A., Hardy, P. H., Cuculis, J. J. & Major, O. H. G. (1970). The unlabelled antibody enzyme method of immunohistochemistry. Preparation and properties of soluble antigen-antibody complex (Horseradish peroxidase-anti-horseradish peroxidase) and its use in identification of Spirochetes. *The Journal of Histochemistry and Cytochemistry*, **18**, 315–33.

Towbin, H., Staehelin, T. & Gordon, J. (1979). Electrophoretic transfer of proteins from polyacrylamide gels to nitrocellulose sheets: procedure and some applicatios. *Proceedings of the National Academy of Sciences, U.S.A.*, **76**, 4350–4.

Trayer, I. P. (1978). Affinity chromatography. In *Techniques in protein and enzyme biochemistry*, Vol. B 1/I, part I, B **102**, ed. H. L. Kornberg, J. C. Metcalfe, D. H. Northcote, C. I. Pogson & K. F. Tipton, pp. 1–34. Amsterdam, Oxford, New York: Elsevier/North-Holland.

Valentines, R. C. (1961). Contrast enhancement in the electron microscopy of viruses. *Advances in Virus Research*, **8**, 287–318.

Van Driessche, E., Smets, G., Dejaejere, R. & Kanarek, L. (1981). The immunohistochemical localization of lectin in pea seeds. *Planta*, **153**, 287–96.

Van Steveninck, R. F. M. & Van Steveninck, M. E. (1978). Ion localization. Techniques based on freeze substitution. In *Electron Microscopy and Cytochemistry of Plant Cells*, ed. J.L. Hall, pp. 187–234. Amsterdam, Oxford, New York: Elsevier/North-Holland.

Vidal, J., Godbillon, G. & Gadal, P. (1980). Recovery of active-highly purified phosphoenolpyruvate carboxylase from specific immunoadsorbent column. *FEBS Letters*, **118**, 31–4.

Vidal, J., Godbillon, G. & Gadal, P. (1983). Influence of light on phosphoenolpyruvate carboxylase in Sorghum leaves. II. Immunochemical study. *Physiologia Plantarum*, **57**, 124–8.

Vithanage, H. I. M. V., Howlett, B. J. & Knox, R. B. (1980). Localization of grass pollen allergen by immunocytochemistry. *Micron*, **11**, 411–12.

Vreeland, V. (1970). Localization of a cell wall polysaccharide in a brown alga with labelled antibody. *The Journal of Histochemistry and Cytochemistry*, **18**, 371–3.

Vreeland, V. (1972). Immunocytochemical localization of the extracellular polysaccharide alginic acid in the brown seaweed *Fucus distichus*. *The Journal of Histochemistry and Cytochemistry*, **20**, 358–67.

Wallace, W. (1977). Proteolytic inactivation of enzymes. In *Regulation of enzyme synthesis and activity in higher plants*, ed. H. Smith, pp. 177–95. London: Academic Press.

Whitehouse, R. L., Benichou, J. C., Couture-Tosi, E., Schenkman, S. & Ryter, A. (1984). Immunolabelling of Bacteriophage receptor protein (LamB) on thin sections of *E. coli* embedded in Lowicryl. *Biology of the Cell*, **51**, 389–94.

Wick, S. M., Seagull, R. W., Osborn, M., Weber, K. & Gunning, B. E. S. (1981). Immunofluorescence microscopy of organized microtubule arrays in structurally stabilized meristematic plant cells. *The Journal of Cell Biology*, **89**, 685–90.

Willingham, M. C. & Yamada, S. S. (1979). Development of a new primary fixative for electron microscopic immunocytochemical localization of intracellular antigens in cultured cells. *The Journal of Histochemistry and Cytochemistry*, **27**, 947–60.

Wrigley, C. W. (1971). Gel electrofocussing. In *Methods in Enzymology*, vol. **22**, ed. S. P. Colowick & N. O. Kaplan, pp. 559–64. New York, London: Academic Press.

Yamada, S. S., Yamada, K. M. & Willingham, M. C. (1980). Intracellular localization of fibronectin by immunoelectron microscopy. *The Journal of Histochemistry and Cytochemistry*, **28**, 953–60.

K. ROBERTS

Antibodies and the plant cell surface: practical approaches

Introduction

Compared with the history of comparable work on the animal cell surface, the enormous power of antibodies to probe components of the cell surface in plants has been recognized only relatively recently. The reasons for this are rather obscure, but three relatively trivial factors at least probably played a role. First, people who worked with plant cells were generally unfamiliar with the process of raising antibodies and often did not have easy access to animals in which to raise the antibodies. Second, many cell wall biochemists were late to realize that oligosaccharides were antigenic and that useful antibodies to them were readily obtainable and last, immunofluorescent methods of staining the plant cells, their plasma membranes and associated cytoplasmic components, such as the cytoskeleton, were very late arriving on the scene (see Doonan & Clayton, this volume). However, progress has been made and this chapter will try to cover the recent advances that have been made in the use of antibodies to study the various components of the plant cell surface. I shall try to put the emphasis on structural questions, and in particular on the practical use of both antisera and monoclonal antibodies (McAb) in pinpointing the location of specific macromolecules at both the light and electron microscope levels. Some of the background to this approach and some detailed examples can be found in a recent companion volume (Roberts *et al.*, 1985*d*) and several chapters in this volume are also relevant (Brewin *et al*; Doonan & Clayton). I shall, in addition, try to draw attention to the value of fluorescent probes other than antibodies, which provide us with useful additional structural data.

The plant cell surface is not a simple concept, but I have chosen to use it deliberately over such phrases as "the plant cell wall" or the "extracellular matrix". So many vital plant cell functions are mediated at the interface between a plant cell and its environment that I believe it makes better biological sense to use a term that includes not only that assemblage of molecules we usually call the cell wall, but also the plasma membrane and that complex assembly of associated submembrane macromolecules which include microtubules and which are referred to as the cortical cytoskeleton. This concept is elaborated more fully in the papers in Roberts *et al.* (1985*c*).

The plant cell surface, as defined above, contains a very large number of macromolecules all of which are potentially antigenic. These include the cell wall polysaccharides, glycoproteins and enzymes, the integral and extrinsic plasma membrane proteins and the cortical microtubules and their associated proteins. Antisera to examples of most of these classes of molecules have been raised by various groups in recent years and have provided information about their structure, location and function, but it seems very likely that their full potential is far from being realized. The sorts of problem for which antibodies are now being used include the identification of oligosaccharide moieties involved in cell signalling and recognition, the mapping of structural domains within cell wall glycoproteins and the cloning of genes coding for hydroxyproline rich structural glycoproteins and incompatibility glycoproteins. Some of these examples, together with others, are discussed more fully in the subsequent sections.

Immunocytochemical approaches

There is a lot of attention focussed at present on the significance of the fact that the plant cell surface contains such a large number of sugars in such a bewildering series of linkages. It increasingly appears, in many cases, that oligosaccharide components of complex glycoproteins and polysaccharides may perform important biological roles as signalling or regulatory molecules (for reviews see Albersheim *et al.*, 1983; Darvill *et al.*, 1985) and it seems certain that antibodies will turn out to be very useful probes for such molecules. Antibodies against carbohydrate antigens have been known for a very long time (see Heidelberger and Bernheimer, 1984) but only recently have they been used in higher plant research. The maximum size for an oligosaccharide determinant is probably about six sugar residues (Kabat, 1966) although in some cases it may be much lower than that (Hamada, Nakajima & Matsuda, 1981; Young *et al.*, 1983; Ashford *et al.*, 1982; Roy, 1984; Anderson, Sandrin & Clarke, 1984; Clarke *et al.*, 1985).

Several groups have now raised antisera against cell wall components, both structural and non-structural, and in particular these have been used to identify cell surface glycoproteins involved in recognition. Using style exudates attempts have been made to identify immunologically the products of incompatibility genes in a variety of species (Ferrari, Bruns & Wallace, 1981; Mau, Raff & Clarke, 1982; Gleeson & Clarke, 1980; Clarke *et al.*, 1985). A potentially very important application of antibodies, which has been little used at present, is to identify antigens that are unique to a particular cell type or tissue type. There are clear advantages in being able to identify rapidly cell types at a very early stage, for example cells in a culture which are capable of becoming embryogenic, and one of the few examples of work in this area is at least encouraging. Raff & Clarke (1981) identified tissue specific antigens that were secreted by suspension culture cells of *Prunus*, including one that was specific to leaf tissue.

Progress in understanding the nature of antigenic determinants (Sutcliffe *et al.*, 1983) has opened the way to making antibodies that are specific for particular sequence determinants. Using such anti-peptide antibodies it has been found that their reactivity is a function of the relative atomic mobility of the sites in the native protein. The more mobile the site the stronger the binding (Tainer *et al.*, 1984). It is quite possible that similar conditions also determine the relative reactivities of anti-carbohydrate antibodies.

In the last few years, probably the greatest advance, in practical terms, in our understanding of the cell wall and its ordered growth and assembly, has been the ability to stain the various components of the cytoskeleton in whole intact cells by indirect immunofluorescence. This has meant that we are no longer dependent on the time consuming and difficult business of 3-D reconstruction of microtubule (MT) arrays from electron micrographs of thin sections. Instead the entire microtubule array can be visualized, not only in individual cells (see Doonan & Clayton, this volume) but also in sheets or groups of cells (Roberts *et al.*, 1985*d*). These advances have all depended on the use of wall degrading enzymes that have loosened the wall texture of paraformaldehyde fixed cells sufficiently to allow the ingress of appropriate antibodies, for example a monoclonal anti-yeast α-tubulin (Kilmartin, Wright & Milstein, 1982) followed by a FITC labelled goat anti-rat IgG antibody. Using these methods a considerable body of data have been generated, not only on the arrangement of MTs during the cell cycle (Wick & Duniec, 1984; Wick *et al.*, 1981; Clayton, 1985; Clayton, Black & Lloyd, 1985; Tiwan *et al.*, 1984; Gunning & Wick, 1985) but also on their relationship to other molecules such as calmodulin (Wick, Muto & Duniec, 1985), actin (Clayton & Lloyd, 1985) and intermediate filament antigens (Dawson *et al.*, 1985). An important feature of these studies using indirect immunofluorescence is that our understanding of the relationship between the cortical microtubule array and the oriented deposition of cellulose in the wall has been dramatically altered. The new emerging picture is of a dynamic integral microtubule array, forming a multistart helix of variable pitch, that can determine the orientation of cellulose deposition in a continuously changing manner. This dynamic helical model (Lloyd, 1983, 1984, 1986; Lloyd & Seagull, 1985) has many attractions and provides an explanatory context for the rapid and dramatic MT and cellulose reorientations that are observed during ethylene treatment (Roberts, Lloyd & Roberts, 1985) and during xylogenesis (Roberts, Burgess & Linstead, 1985; Falconer & Seagull, 1985). Although the methods for simple immunofluorescence staining are given elsewhere in this volume (see Doonan & Clayton; Perrot-Rechenmann & Gadal), it is appropriate at this point to discuss some of the principles involved in observing the resultant fluorescent specimen. Details of the use of specific filters are included in a later section.

Fluorescence microscopy

The use of fluorescence in plant cell microscopy has a long history (O'Brien & McCully, 1981; Holz, 1974; Jeffree, Yeoman & Kilpatrick, 1982), but it is only relatively recently that the power of epifluorescent methods, coupled with fluorescently labelled antibodies, has really been widely realized. This has been partly to do with the evolution of new equipment, partly to do with the previous lack of appropriate antibodies or probes (Knox & Clarke, 1978), and partly to do with the slow introduction in plants of the suitable staining methods mentioned earlier (Simmonds, Seagull & Setterfield, 1985).

The most commonly used fluorochromes in immunofluorescence studies are fluorescein isothiocyanate (FITC) and rhodamine. These are intense fluorescent chromophores that are excited by blue and green light respectively which, when excited, emit light in the green and red wavelengths. These non-overlapping excitation and emission spectra mean that with the appropriate filters (see last section) the two stains can be used together to reveal different cell components such as actin and microtubules (Clayton & Lloyd, 1985). In addition, other fluorescence markers that are activated by UV wavelengths can be used to identify the state of chromosome condensation for example (Clayton, 1985), or the sites of cell wall synthesis (Roberts *et al.*, 1985*a*; see Fig. 1). This means that, with narrow band pass filters, at least three separate localizations can be made on the same cell with fluorescent techniques. It is worth emphasizing here a few points about the intensity of fluorescence that can be seen by incident light excitation (epifluorescence). It is often forgotten that many simple factors can dramatically reduce the potential signal. Quenching can occur by a variety of chemicals such as oxygen or ions such as iodide, and even by an excess of the fluorochrome itself. Emission for each specific fluorochrome is also strongly dependent on the pH of the mountant. For example, FITC has an optimum at about pH 9.0. The light path from source to specimen should be as short as possible, as the intensity falls off inversely with the square of the distance. It is worth remembering also that for most fluoro-

Fig. 1. Various stages of cell division in root tip meristem cells of maize. Four cells are shown and in each pair, the image on the left is photographed with FITC filters to reveal the position of the microtubules by immunofluorescence (see Doonan & Clayton, this volume) and the image on the right is the same cell photographed using a DAPI filter to show the behaviour of the chromosomes and nucleus. Fig. 1a and b show a cell in prophase with a clear preprophase band of microtubules and a spindle forming. Fig. 1c and d show a cell in early anaphase with two clear half spindles and separating groups of chromosomes. Fig. 1e and f show a cell in telophase with two daughter groups of chromosomes separated by a prominent cell plate and phragmoplast microtubules. Fig. 1g and h show a cell in late cytokinesis in which the two daughter nuclei are reforming and the cell plate with its marginal phragmoplast microtubules has nearly reached the mother cell wall.

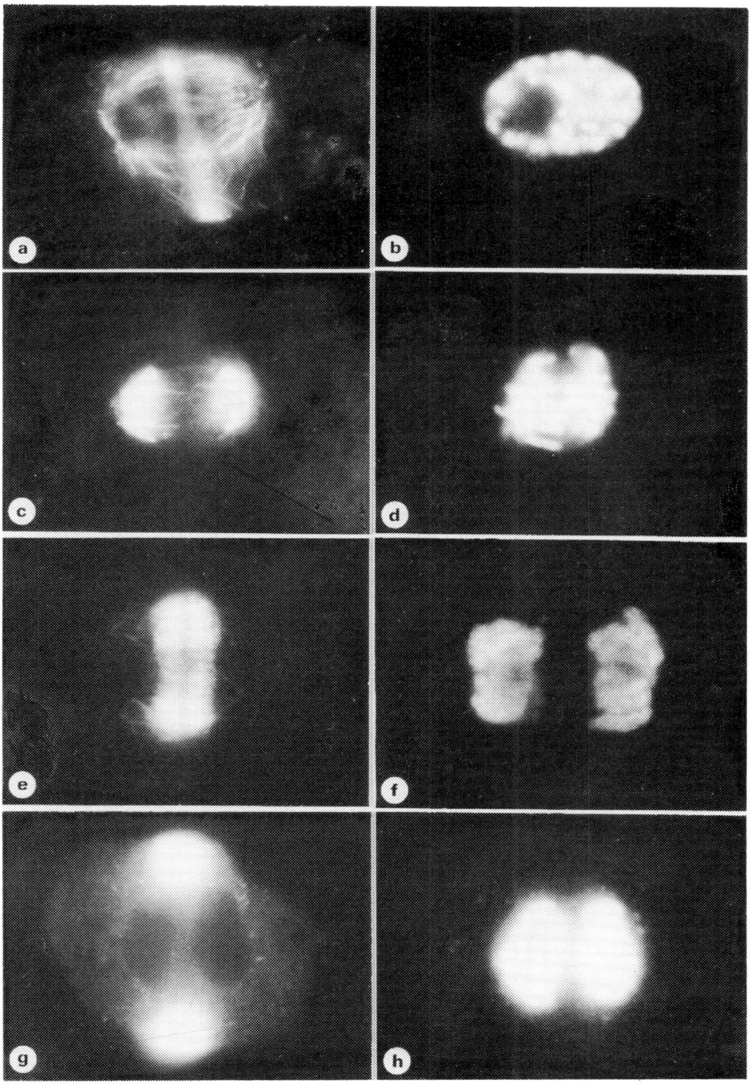

chromes (e.g. FITC and rhodamine) a mercury or UV source is entirely unnecessary. Indeed the FITC excitation maximum does not correspond to any of the sharp emission lines from a mercury source and a quartz iodine, xenon or even an ordinary tungsten lamp may be preferable (O'Brien & McCully, 1981; Holz, 1974). Because the objective in epifluorescence acts as both the condenser and the objective, the intensity of signal transmitted depends on the fourth power of the numerical aperture (NA) of the lens. The considerable gain in brightness using high aperture objectives, however, must be balanced against the consequent loss of depth of field. Thus, a Plan-Neofluor X63 NA 1.25 will produce an image less than two thirds of the brightness of a Plan-Apo X63 NA 1.4 but will have a much better depth of field. In the case of some manufacturers (e.g. Zeiss) the flat field, highly corrected Plan Apochromats recommended for fluorescence work (in particular the X40 NA 1.0 and the X63 NA 1.4) do not unfortunately transmit light in the UV, making them unsuitable for e.g. DAPI staining of chromosomes, Calcofluor white staining for cellulose or Aniline Blue staining for callose. On balance, I find that the most useful lenses are the UV transmitting flat field Neofluor objectives such as the Zeiss Plan-Neofluors and the Leitz NPL Fluotars. With increasingly sensitive methods of recording fluorescent images, such as presensitized film or image intensified video cameras, the need for the highest aperture lenses is diminished and the technique is now routinely capable of detecting, literally, a few molecules per μm^2. This becomes very important in experiments using fluorescent DNA probes to locate and map single gene copies in a chromosome set. The rapid fading of most fluorochromes can also be a problem but the use of new mountants (Johnson *et al.*, 1982) has helped considerably in this respect. Other, potentially important, developments have not yet found users in plant cell research. Thus the ability to do immunofluorescence on epon embedded material means that correlative observations can be made on the same section in the light and electron microscope (Rider, Nuss & Bowser, 1984). It may even be possible to label with both fluorescent and gold labelled second antibodies. Another recent development is immunophotoelectron microscopy (Birrell *et al.*, 1985) in which animal cytoskeletons have been revealed with remarkable clarity using the electron optical analogue of immunofluorescence. In animal cells it has been found that the uncoupled fluorochromes themselves can act as remarkable localization reagents. Thus, Johnson, *et al.* (1980) showed that rhodamine 123 can specifically localize mitochondria in living cells, as revealed by fluorescence microscopy. Our initial experiments with this dye, however, suggest it may not be taken up by plant cells in quite the same way.

Immunogold methods

Practical aspects of applying immunocytochemistry at the level of the electron microscope are covered elsewhere in this volume (see Brewin *et al.*; Doonan & Clayton; Perrot-Rechenmann & Gadal) and I shall confine myself here to other aspects of immunogold labelling and the cell surface. Good recent reviews of colloidal gold markers can be found in De Mey (1983, 1984), Bendayan (1984) and Roth (1983). Surprisingly, considering the many advantages, the fact that colloidal gold can be visualized in the light microscope as a red marker, as well as in the electron microscope as an electron dense probe, has not been much exploited. An exception is the very elegant study (De Mey *et al.*, 1981) of the behaviour of microtubules during cell division in the endosperm of *Haemanthus*. This, however, is a rather special case as the cells examined do not have cell walls present. How applicable the technique is in walled cells remains to be seen. The use of labelled frozen sections of fixed tissue, that are then embedded and thin sectioned, is a possible approach which has already produced results (Raikhel, Mishkind & Palevitz, 1984).

Although the colloidal gold probe is usually coupled either to a second antibody or to protein A it is, of course, perfectly possible to couple it to any molecule that is involved in specific interactions with another molecule. Thus, colloidal gold has been used in conjunction with lectins, to probe the location of specific saccharide residues and also in conjunction with enzymes to locate their substrates. This latter approach, which has great potential for studies of plant cell wall architecture, has been elegantly pioneered by Vian, Brillouet & Satiat-Jeunemaitre (1983) in which they used xylanase-gold complexes to visualize xylans in the secondary cell walls of hardwood tissue.

With all the gold probes mentioned there are essentially two strategies for staining thin sections. In the first, potential non-specific binding is prevented by blocking such sites with another protein, usually bovine serum albumin or low fat dried milk. In the second, no specific blocking agent is used and the background labelling is reduced by increasing the stringency of binding, usually by including a dilute detergent in the washes. In general, we have found that the latter method (Beesley *et al.*, 1984) produces less gold grains per known numbers of antigen molecules, but that this is more than offset by the very much lower background labelling. The importance of adequate controls and an understanding of quantitative and statistical approaches to gold labelling should be emphasized and are well illustrated in an elegant study of constituents in the legume-*Rhizobium* symbiosis (Brewin et al., 1985; Robertson *et al.*, 1984, 1985). This emphasizes the need for intensive characterization of the antibodies used for immunogold studies. Even minor cross-reactivities, with antigens other than that of interest, can lead to confusing or misleading results. This is particularly true of a cell surface molecule which may contain oligosaccharide antigens present on several other molecules. Whether antisera are raised against a purified antigen such as a band off a gel (Roberts *et al.*, 1985*d*) or are of monoclonal origin, this problem can still be present (Roberts *et al.*, 1985*b*). Western blotting, preferably of 2-dimensional gels is the usual test for monospecificity (see Cuming *et al.*, this volume), but even here it is feasible for an antibody to label fixed cell components by immunocytochemistry but not to label the corresponding antigen on a Western blot. The obverse, i.e. the reduction of antigenicity by tissue preparation methods, can also be observed (Craig & Goodchild, 1982).

A major improvement in the level of gold labelling has resulted from the use of embedding resins at low temperatures. Lowicryl K4M is now a routine and very successful example of a low temperature resin (Wells, 1985) and recent applications from our own laboratory include the localization of spiralin in *Spiroplasma* cells (Townsend & Plaskitt, 1985), the spatial distribution of various chloroplast components (Shaw & Henwood, 1985) and the localization of components in the legume-*Rhizobium* symbiosis (Robertson *et al.*, 1985). A useful technical point that has emerged during a study of the binding of anti-tubulin antibodies to *Chlamydomonas* flagella (Wells & Roberts, manuscript in preparation) is that by supporting the section over a thin carbon net it is possible to not only double the labelling efficiency on the unsupported portions of the section (both sides of the section then being available) but to decrease the background considerably. This suggests that the carbon support film itself may be a major cause of non-specific gold grains binding.

Monoclonal antibodies

The application of hybridoma technology to plant cell problems has very recently begun to blossom. This is not surprising when one considers the unique specificities offered by monoclonal antibodies (McAb), and the enormous range of important questions in plant cell biology that they can help to answer. It is equally not surprising that most of the examples published to date are in the area of the cell surface. The role of the cell surface in signalling and recognition events and in the control of normal growth and development hardly needs emphasizing (Roberts *et al.*, 1985*c*) and the problem of identifying and characterizing the key surface macromolecules involved in these processes is an ideal one for McAb to solve.

There are now several examples of this approach in lower plant forms.

For example, in *Dictyostelium*, monoclonals have been used to identify a plasma membrane glycoprotein implicated in cell-cell adhesion (Murray, Ninnan & Loomis, 1983) and to characterize the extracellular slime sheath (Grant & Williams, 1983).

In *Phytophthora* monoclonals have been raised to the surface of zoospores and cysts (A.R. Hardham, E. Suzaki and J. Perkin, personal communication), in *Fucus* developmentally regulated cell wall carbohydrate antigens have been located using McAb (Vreeland & Laetsch, 1984) and the monoclonals to antigens in the peribacteroid membrane from *Rhizobium*-induced root nodules in peas have already been mentioned (see Brewin *et al.*, this volume).

In higher plants the emphasis has been on cell surface molecules involved in signalling and recognition. Thus plasma membrane antigens involved in high affinity binding of abscisic acid have been identified in bean (Hornberg & Weiler, 1984) and a plasma membrane antigen involved in auxin transport has been shown in peas (Jacobs & Gilbert, 1983). In the latter case immunofluorescence showed the transport protein to be specifically localized at the basal end of the pea hypocotyl cells.

Perhaps the largest collection of McAb to plant antigens can be found in the related work of at least five groups investigating the cell surface of members of the Volvocales. At present little comparative work has been done, but the collection of antibodies as a whole opens up the potential for a detailed characterization of the extracellular matrix in *Chlamydomonas* and its near relatives. Adair (1985) has raised a library of monoclonals to the sexual agglutinins of *C. reinhardi*, groups of which bind to three different functional domains of the large glycoprotein molecule, and some of which also bind to cell wall glycoproteins. W.L. Homan *et al.* (personal communication) have raised a similar family of McAb to sexual agglutinins, but in the related species *C. eugametos*. Woodward, Young & Bloodgood (1985), using *C. reinhardi*, have made McAb to carbohydrate epitopes on the flagella surface glycoproteins and glycolipids and shown again that cross reactivity to a range of other cell surface molecules is common.

Kirk's group (D. Kirk, personal communication; Hoffmann, 1984) have raised a library of McAb to the extracellular matrix glycoproteins of the multicellular *Volvox* and have demonstrated not only cross reactivity with unicellular ancestors but developmental regulation of some of the antigens. And lastly, Smith *et al.* (1984*a*, 1984*b*) raised a family of 20 McAb to the major structural glycoprotein of the cell wall of *C. reinhardi*. Some of these show cross reactivity with flagellar surface antigens and have been used in both immunofluorescence and immunogold localization studies (Roberts *et al.*, 1985*d*, 1985*b*; C. Grief, manuscript in preparation). One of the important steps in both the screening of McAb and in the detailed characterization of antigens, is the quick and sensitive detection of positive binding between the antibody and the antigen. This can be complicated if the antigen is in short supply or cannot be readily purified. We have found two methods, that were generated during the course of producing antibodies to the major structural glycoprotein in *C. reinhardi* (Smith *et al.*, 1984*a*), to be very widely applicable. Both methods depend on the immobilization of antigens, native or modified, to nitrocellulose sheets, followed by their detection with antibodies and enzyme linked second antibodies. The first, useful for primary screening, uses very small dots of hybridoma culture supernatant applied directly to a small antigen dot on nitrocellulose. If longer incubations were needed, or competition assays required, a manifold was used to allow an increase in the volume of antiserum present. The detailed method is given in the Protocols. The second method, useful in later screening, is designed to probe Western blots of antigens with different antisera without cutting up the sheet of nitrocellulose. A metal template is clamped on either side of the nitrocellulose blot and antisera is run into the channels that are formed. Details of this method are given in Smith *et al.* (1984*b*).

Other fluorescent probes

Although this chapter is primarily about immunological approaches to the plant cell surface, there are other fluorescent probes useful in structural investigations that it is important to mention in the context of plant cell research. Some of these probes are particularly useful in conjunction with fluorescent antibody studies.

1. Lectins

Fluorescently tagged lectins have been used in many studies of plant cell behaviour, particularly in following the presence and function of various glycoconjugates. FITC-conjugated lectins, with different sugar binding sites, have been used to investigate the molecular architecture of the cell surface in a wide variety of algal species (Sengbusch *et al.*, 1982; Sengbusch & Muller, 1983) and aspects of the developmental regulation of the sugar ligands were revealed. Similar events in the sexual cycle and development of both *Chlamydomonas* (Millikin & Weiss, 1984) and *Volvox* (Kurn, 1981) have also been demonstrated using Concanavalin A. As a general point, however, none of the available lectins have the unique range of specificities that are available with antibodies and consequently are unlikely to provide us with such precise information.

2. Aniline blue

Under suitable conditions this dye can be used as a fluorescent marker for ß1–3 linked D-glucans (callose). Sirofluor, the fluorochrome present in aniline blue, has now been isolated and characterized (Evans, Hoyne & Stone, 1984; Stone *et al.*, 1984) and in aqueous solutions at pH 8.5 it absorbs in the ultraviolet (380–95 nm) and produces a yellow fluorescent emission (480–95 nm). This stain has great potential for studying early events in wounding and pathogen attack when callose is an early visible marker. (Hinch *et al.*, 1985; Wetherbee *et al.*, 1985), but has not as yet been used in conjunction with other fluorochromes in immunofluorescent studies. Manufacturers might note that a narrow band pass filter set does not at present appear to be available for specific Sirofluor fluorescence and that this fact might be preventing its more widespread use.

3. Calcofluor white

This optical brightener has been used in a variety of studies on cellulose biosynthesis (Roberts *et al.*, 1982). The dye binds tightly to nascent cellulose chains and can be seen by its intense blue fluorescence when illuminated with UV light. Luckily, the specific filter set used for DAPI and Hoechst staining of DNA is ideally suited for this fluorochrome (see below) and this allows its use in conjunction with FITC labelled antibodies. For example, in studies of tracheary element differentiation in *Zinnia*, Calcofluor white was used, at a concentration of 0.01%, to show the timing of ordered cellulose deposition in relation to the rearrangements of the cortical microtubules (Roberts *et al.*, 1985*a*) revealed by FITC anti-tubulin fluorescence (Fig. 2). Similar results have been obtained by Falconer & Seagull (1985). In the former study, the fact that lignin autofluoresces could be used to compare the location and timing of lignin deposition in the cell wall thickenings in the same cells as microtubule rearrangements and cellulose deposition were monitored. Another elegant use of Calcofluor has been as a "pulse label" in studying cell elongation in red algae. The nascent cellulose is vitally stained, the dye washed out, and the position of the fluorescence monitored as the cells continue to grow (Waaland & Waaland, 1975).

4. DAPI

DAPI, 4′6-diamidino-2-phenylindole, like Hoechst, is a specific probe for DNA and can be used to show it in chromatin, mitochondria and chloroplasts (James & Jope, 1978). At a concentration of 1 μg cm^{-3} in the final wash, it makes a very useful "counterstain" to FITC immunofluorescence. The stain is not fluorescent unless bound to DNA. For example the behaviour of the chromosomes and of the microtubules can be seen with non-overlapping fluorescence. DAPI absorbs in the UV (360 nm) and emits in the blue. A set of specific filters (see below) rather than the general UV filter set usually supplied is needed to separate the fluorescence from FITC and rhodamine. Specific filter sets are particularly important in working with plant material because of interfering autofluorescence from cell walls and chloroplasts. The combination of DAPI and anti-tubulin

Fig. 2. *Zinnia* mesophyll cells undergoing *in vitro* tracheary element formation. Fig. 2a and b show the same pair of cells stained for microtubules (a) and with Calcofluor white (b). The ordered groups of microtubules, which anticipate the position of cellulose cell wall thickenings, can be seen in both cells but only one (b) has started cellulose deposition. Fig. 2c and d show two cells which have undergone extensive cellulose cell wall thickenings. The photographs are taken using the DAPI filter set and Calcofluor white induced fluorescence. The pictures are from unpublished data of K. Roberts, P. Linstead and J. Burgess.

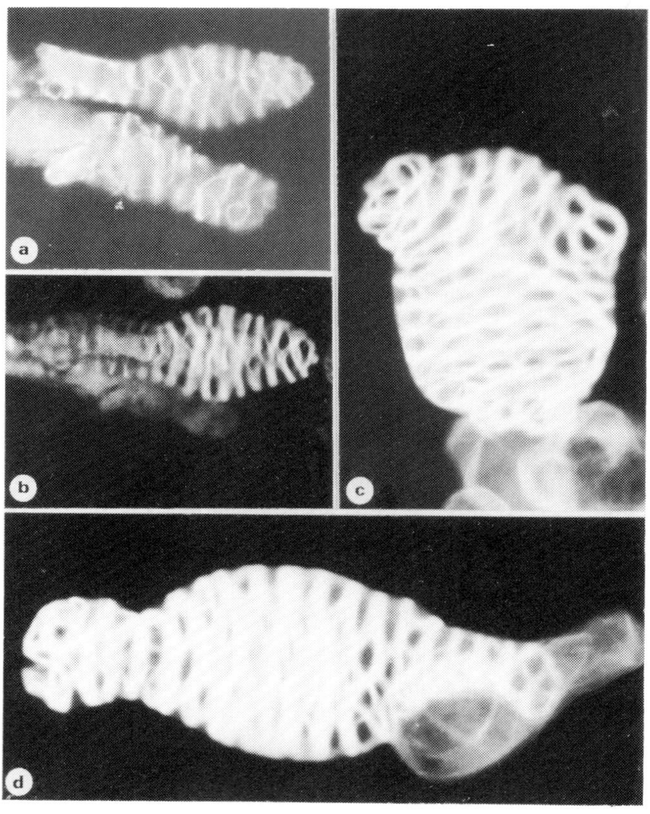

antibody staining has been used to study the exact timing of the "prepro-phase band" of microtubules in a variety of cell types (Roberts *et al.*, 1985*a*; L. Clayton, personal communication). In most cases the band of MTs appears at the same time as chromosome condensation begins and peri-nuclear tubulin fluorescence appears (Fig. 1). It persists until prometa-phase starts. Our conclusion was that as the band is always present throughout the whole of classically defined prophase and only rarely appears before clear chromosome condensation, it is more accurately called a "prophase band" of microtubules.

The use of filters in fluorescence microscopy

It is notorious that plant cells, unlike most animal cells, can display a wide range of natural fluorescence, often creating problems and con-fusion for those trying to use immunofluorescence techniques (O'Brien and McCully, 1981). The use and selection of specific fluorochromes and filter sets for plant cell work is therefore an important question and a few remarks on some of the more useful selective filters are appropriate here. Within the visible light spectrum (450–750 nm) it is now quite feasible to arrange selective filtering to allow at least three different fluorochromes to be separately distinguished. One or more of these colours may be useful to locate organelles, e.g. chloroplast autofluorescence, Calcofluor white for

Fig. 3. Schematic diagram to show the light path in epifluorescence microscopy and the three main filters used. The dichroic beam splitting mirror reflects light down through the objective to the object but transmits emitted fluorescence above its wavelength to the eyepiece.

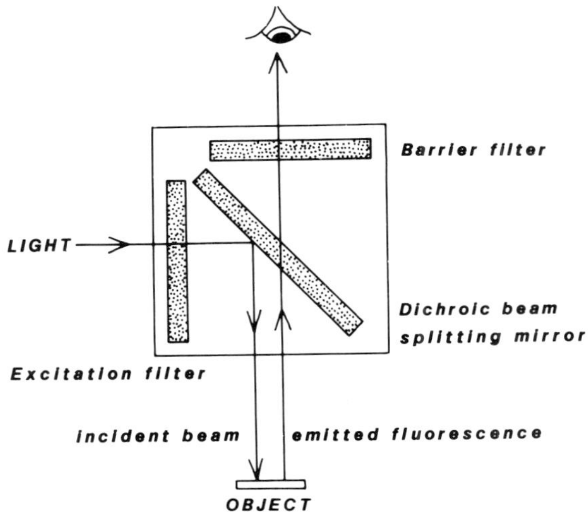

cellulose deposition or DAPI to determine nuclear or organelle location and behaviour while the other(s) may be used for immunofluorescence with, e.g. FITC, Rhodamine or Texas Red. The three filters which comprise the usual incident light or epifluorescence equipment are shown in Fig. 3 and the transmission characteristics of the three most useful and commonly used filter sets in plant cell work are given in Table 1 and Fig. 4. Band pass filters are characterized by either their short and long wave half value width (half-power points) e.g. BP 450–90, or by their centre wavelength i.e. maximum transmission followed by the half-value width (half-power band width) e.g. BP 365/11. The dichromatic beam splitters, and the long and short pass filters are marked with figures representing the cut-off wavelength in nm. Some of the selective filter sets are not generally

Table 1. *Filter sets for fluorescence microscopy.*[1]

	Zeiss filter set		Leitz filter block	
General FITC	9	BP 450–490 FT 510 LP 520	H	BP 390–490 RKP 510 LP 515
Selective FITC[2]	10	BP 450–490 FT 510 BP 520–560	L2	BP 450–490 RKP 510 BP 525/20
Rhodamine	15	BP 546/2 FT 580 LP 590	M2	BP 546/14 RKP 580 LP 580
DAPI/Hoechst[3]	—	BP 365/11 FT 425 BP 450–490	A	BP 340–380 RKP 400 LP 430

Note: The table shows the characteristics of the filter sets from two selected manufacturers, for FITC, rhodamine and DAPI fluorescence, the three most useful and commonly used non-overlapping emitters in plant cell immunofluorescence work. Comparable filter sets are available or can be made up by most major microscope manufacturers. The figures given are for the exciter filter, the dichromatic beam splitter and the barrier filter respectively (see Fig. 3). For an illustration of the spectral features of the three filter sets see Fig. 4. [1]Filter abbreviations: BP – band pass filter; RKP or FT – dichroic beam splitting mirror; LP – long pass filter; KP – short pass filter. Numbers indicate wavelengths in nm. [2] It is a more flexible arrangement for FITC fluorescence to use the general FITC filters together with a separate removable red barrier filter (e.g. KP 560). The net result is the same as the selective filter set, but with plant cells it has the advantage that, by removing the red filter, chloroplast location, for example, can be seen by the chlorophyll autofluorescence (see Fig. 5). [3]The Zeiss selective DAPI set is non-standard but can be made up by them. The Leitz DAPI set is a general UV filter set and will let through all wavelengths above 430 nm.

available yet, but all major manufacturers will assemble them to a customer's requirements e.g. the selective DAPI filter sets. It is worth reiterating here that confusion can sometimes be caused by the fact that some planapochromat objectives (e.g. Zeiss) do not transmit light in the UV range and that care in the choice of objectives, therefore, is often necessary.

Fig. 4. Transmission spectra of three selective filter sets to show how they can be used to detect FITC, rhodamine and DAPI fluorescence without mutual interference. (Figures compiled from data published by C. Zeiss, West Germany.)

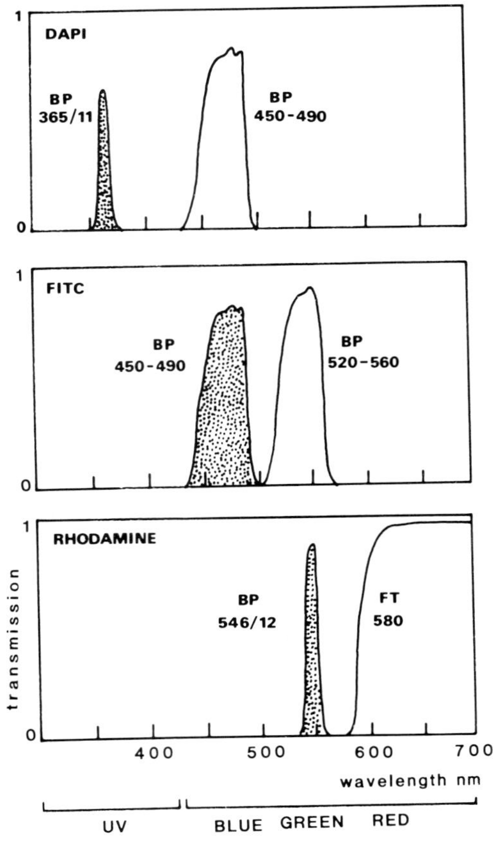

Fig. 5. Diagram to show how long a pass filter (LP 520), which lets through all fluorescence emitted from green through to red (e.g. both FITC and chlorophyll), can be converted to a selective FITC filter by adding a removable red barrier filter (KP 560). The net result of two barrier filters together is equivalent to a band pass, BP 520–560, filter (Fig. 4).

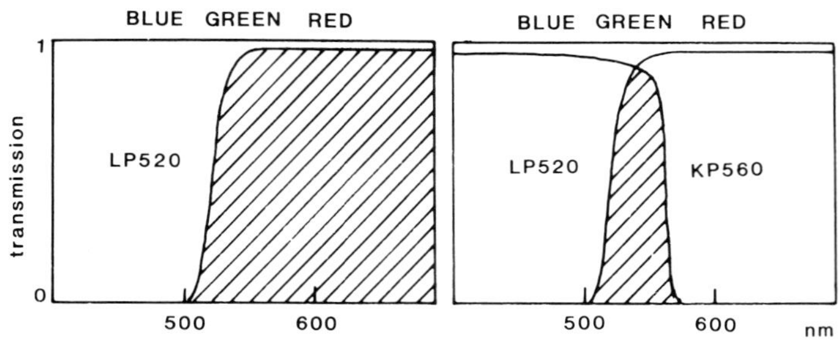

Fig. 6. An example of the use of the dot-binding assay in screening hybridoma supernatants. In this case, 1 μl dots of antigen (containing a cell wall glycoprotein) were assayed directly with 2 μl aliquots of supernatants from wells containing growing cells. Numerous positive monoclonal antibodies can be seen e.g. A9, A10 together with some negative wells e.g. H3, H5 (from Smith *et al.*, 1984*b*),

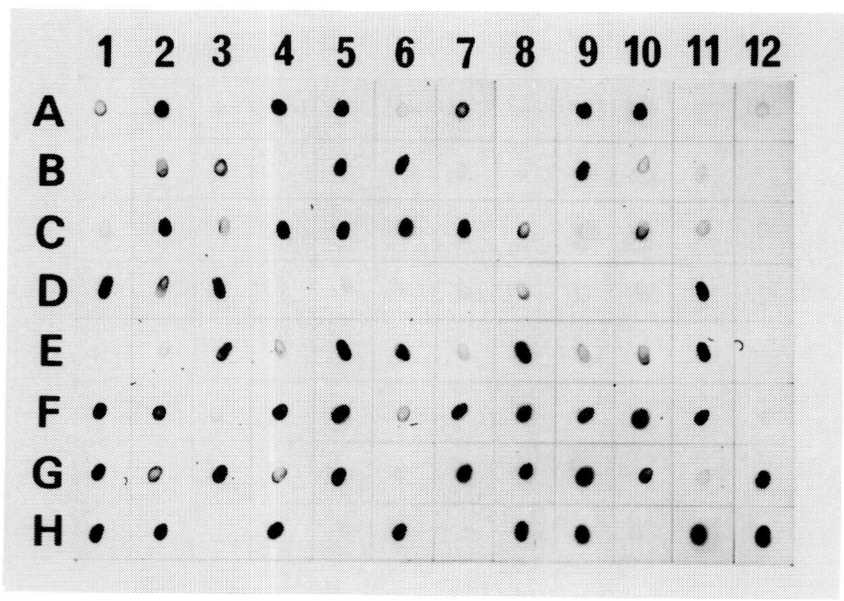

PROTOCOLS

I. Dot immunoblotting – rapid screening for monoclonal antibodies (see Smith et al., 1984a)

1. Rule a nitrocellulose sheet in 9 mm squares using a paper template on a light box[a].
2. Wash the sheet for 5 min in distilled water and dry for at least 1 h at room temperature.
3. Apply 1 μl dots of a 1 mg cm^{-3} solution of antigen to the centre of each square, with sheet on a light box[b].
4. Air dry the sheet[c].
5. Block the sheet with 20 cm^3 of 1–3% BSA or 3% low fat dried milk in TBS[d] in the lid or container from 96 well microtitre plates on a rotary shaker[e].
6. Rinse in TBS and dry.
7. Apply 1.5 or 2.0 μl samples of hybridoma culture supernatant directly onto antigen spots and leave for 5 min.
8. Wash for 30 min in several changes of TBS.
9. Replace in the blocking solution retained from step 5.
10. Incubate in above solution, to which is now added horseradish peroxidase-conjugated anti-mouse immunoglobulin[f] for 2 h.
11. Wash for 30 min in several changes of TBS.
12. Develop by incubating in 5 vols TBS, 1 vol 3 mg cm^{-3} freshly prepared 4-chloro-1-naphthol in methanol made 0.01% (v/v) with hydrogen peroxide[g].
13. Wash sheet in distilled water, dry in the dark and store sealed in polythene in the dark.

Notes

a Wear gloves at all times when handling to avoid damage to the sheet.
b For convenience a Hamilton syringe fitted with a multidispenser attachment is quite useful. With a relatively pure antigen, 0.1 μg cm^{-3} is sufficient, but for a more impure or unknown antigen a more suitable concentration might be 1 mg cm^{-3}.
c Sheet may be stored at 4 °C at this point.
d TBS: tris buffered saline – 50 mM tris, 200 mM NaCl, pH 7.4.
e Retain the blocking solution for step 9.
f If the McAb are raised in rats then use the appropriate anti-rat IgG. The final concentration of this second antibody may be varied, but a 1:1000 dilution is a good starting point.
g The colour (blue grey) develops in 5 min (see Fig. 6).

KEY REFERENCES

Brewin, N. J., Robertson, J. G., Wood, E. A., Wells, B., Larkins, A. P., Galfre, G. & Butcher, G. W. (1985). Monoclonal antibodies to antigens in the peribacteroid membrane from *Rhizobium*-induced root nodules of pea cross-react with plasma membranes and Golgi bodies. *EMBO Journal*, **4**, 605–11.

Kohler, G. (1985). Derivation and diversification of monoclonal antibodies. *EMBO Journal*, **4**, 1359–65.

Lloyd, C. W. (1984). Towards a dynamic helical model for the influence of microtubules on wall patterns in plants. *International Reviews in Cytology*, **86**, 1–51.

Roberts, K., Lloyd, C. W., Shaw, P., Johnston, A. W. B. & Woolhouse, H. W., eds (1985*c*). The cell surface in plant growth and development. *Journal of Cell Science*, Supplement 2.

Roberts, K., Phillips, J., Shaw, P., Grief, C. & Smith, E. (1985*d*). An immunological approach to the plant cell wall. In *The Plant Cell Wall*, Society for Experimental Biology, Seminar Series No. 28, eds. C. T. Brett & J. R. Hillman, pp. 125–54. Cambridge: Cambridge University Press.

REFERENCES

Adair W. S. (1985). Characterization of *Chlamydomonas* sexual agglutinins. *Journal of Cell Science*, Supplement 2, 233–60.

Albersheim, P., Darvill, A. G., McNeil, M., Valent, B. S., Sharp, J. K., Nathnagel, E. A., Davis, K. R., Yamazaki, N., Gollin, D. J., York, W. S., Dudman, W. F., Darvill, J. E. & Dell, A. (1983). Oligosaccharins: naturally occurring carbohydrates with biological regulatory functions. In *Structure and Function of Plant Genomes*, ed. O. Ciferri & L. Dure, pp. 293–312. New York: Plenum Press.

Anderson, M. A., Sandrin, M. S. & Clarke, A. E. (1984). A high proportion of hybridomas raised to a plant extract secrete antibody to arabinose or galactose. *Plant Physiology*, **75**, 1013–16.

Ashford, D., Desai, N. N., Allen, A. K., Neuberger, A., O'Neill, M. A. & Selvendran, R. R. (1982). Structural studies of the carbohydrate moieties of lectins from potato (*Solanum tuberosum*) and thorn apple (*Datura stramonium*) seeds. *Biochemical Journal*, **201**, 199–208.

Beesley, J. E., Day, S. E. J., Betts, M. P. & Thorley, C. (1984). Immunocytochemical labelling of bacteroides nodosus pili using an immunogold technique. *Journal of General Microbiology*, **130**, 1481–7.

Bendayan, M. (1984). Protein A-gold electron microscopic immunocytochemistry: methods, applications, and limitations. *Journal of Electron Microscopical Technology*, **1**, 243–70.

Birrell, A. B., Habliston, D. L., Nadakavukarenke, K. & Griffith, O. H. (1985). Immunophotoelectron microscopy: the electron optical analogue of immunofluorescence microscopy. *Proceedings of the National Academy of Sciences, U.S.A.*, **82**, 109–13.

Brewin, N. J., Robertson, J. G., Wood, E. A., Wells, B., Larkins, A. P., Galfre, G. & Butcher, G. W. (1985). Monoclonal antibodies to antigens in the peribacteroid membrane from *Rhizobium*-induced root nodules of pea cross-react with plasma membranes and Golgi bodies. *EMBO Journal*, **4**, 605–11.

Clarke, A. E., Anderson, M. A., Bacic, T. & Harris, P. J. (1985). Molecular basis of cell recognition during fertilization in higher plants. *Journal of Cell Science*, Supplement 2, 261–85.

Clayton, L. (1985). The cytoskeleton and the plant cell cycle. In *The Cell Division Cycle in Plants*, Society for Experimental Biology Seminar Series 26, eds. J. A. Bryant & D. Francis, pp. 113–31. Cambridge: Cambridge University Press.

Clayton, L., Black, C. M. & Lloyd, C. W. (1985). Microtubule nucleating sites in higher plant cells identified by an autoantibody against pericentriolar material. *Journal of Cell Biology*, **101**, 319–24.

Clayton, L. & Lloyd, C. W. (1985). Actin organization during the cell cycle in meristematic plant cells. Actin is present in the cytokinetic phragmoplast. *Experimental Cell Research*, **156**, 231–8.

Craig, S. & Goodchild, D. J. (1982). Post embedding immunolabelling. Some effects of tissue preparation on the antigenicity of plant proteins. *European Journal of Cell Biology*, **28**, 251–6.

Darvill, A. G., Albersheim, P., McNeil, M., Lau, J. M., York, W. S., Stevenson, T. T., Thomas, J., Doares, S., Gollin, D. J., Chelf, P. & Davis, K. (1985). Structure and function of plant cell wall polysaccharides. *Journal of Cell Science*, Supplement 2, 203–17.

Dawson, P. J., Hulme, J. S. & Lloyd, C. W. (1985). Monoclonal antibody to intermediate filament antigen cross-reacts with higher plant cells. *Journal of Cell Biology*, **100**, 1793–8.

De Mey, J. (1983). Colloidal gold probes in immunocytochemistry. In *Immunocytochemistry: Applications in Pathology and Biology*, eds. J. Polack & S. van Noorden, pp. 82–113. London: J. Wright.

De Mey, J. (1984). Colloidal gold as marker and tracer in light and electron microscopy. *EMSA Bulletin*, 54–66.

De Mey, J., Lambert, A., Bajer, A. S., Moeremans, M. & De Brabander, M. (1981). Visualization of microtubules in interphase and mitotic plant cells of *Haemanthus* endosperm with the immunogold staining method. *Proceedings of the National Academy of Sciences, U.S.A.*, **79**, 1898–902.

Evans, N. A., Hoyne, P. A. & Stone, B. A. (1984). Characteristics and specificity of the interaction of a fluorochrome from aniline blue (Sirofluor) with polysaccharides. *Carbohydrate Polymers*, **4**, 215–30.

Falconer, M. M. & Seagull, R. W. (1985). Immunofluorescent and Calcofluor White staining of developing tracheary elements in *Zinnia elegans* L. suspension cultures. *Protoplasma*, **125**, 190–8.

Ferrari, T. E., Bruns, D. & Wallace, D. H. (1981). Isolation of a plant glycoprotein involved with control of intercellular recognition. *Plant Physiology*, **67**, 270–7.

Gleeson, P. A. & Clarke, A. E. (1980). Antigenic determinants of a plant proteoglycan, the *Gladiolus* style arabinogalactan-protein. *Biochemical Journal*, **191**, 437–47.

Grant, W. N. & Williams, K. L. (1983). Monoclonal antibody characterization of slime sheath: the extracellular matrix of *Dictyostelium discoideum*. *EMBO Journal*, **2**, 935–40.

Gunning, B. E. S. & Wick, S. M. (1985). Preprophase bands, phragmoplasts, and spatial control of cytokinesis. *Journal of Cell Science*, Supplement 2, 157–79.

Hamada, T., Nakajima, T. & Matsuda, K. (1981). Comparison of the mannan structure from the cell-wall mutant *Candida* sp. M-7002 and its wild type. *European Journal of Biochemistry*, **119**, 373–9.

Heidelberger, M. & Bernheimer, A. W. (1984). Cross reactions of polysaccharides of fungi, molds, and yeasts in anti-pneumococcal and other antiserea. *Proceedings of the National Academy of Science, U.S.A.*, **81**, 5247–9.

Hinch, J. M., Wetherbee, R., Mallett, J. E. & Clarke, A. E. (1985). Response of *Zea mays* roots to infection with *Phytophthora cinnamomi*. I. The Epidermal Layer. *Protoplasma*, **126**, 178–87.

Hoffman, J. L. (1984). Monoclonal antibodies to surface components of *Volvox*. *Journal of Cell Biology*, **99**, 243*a*.

Holz, H. M. (1974). *Worthwhile facts about fluorescence microscopy*. West Germany: C. Zeiss.

Hornberg, C. & Weiler, E. A. (1984). High-affinity binding sites for abscisic acid on the plasmalemma of *Vicia faba* guard cells. *Nature*, **310**, 321–4.

Jacobs, M. & Gilbert, S. F. (1983). Polar auxin transport. *Science*, **220**, 1297–300.

James, T. W. & Jope, C. (1978). Visualization by fluorescence of chloroplast DNA in higher plants by means of the DNA-specific probe 4′,6-diamidino-2-phenylindole. *Journal of Cell Biology*, **79**, 623–30.

Jeffree, C. E., Yeoman, M. M. & Kilpatrick, D. C. (1982). Immunofluorescence studies on plant cells. *International Review of Cytology*, **80**, 231–65.

Johnson, L. V., Walsh, M. L. & Chen, L. B. (1980). Localization of mitochondria in living cells with rhodamine 123. *Proceedings of the National Academy of Science, U.S.A.*, **77**, 990–4.

Johnson, G. D., Davidson, R. S., McNamee, K. C., Russell, G., Goodwin, D. & Holborow, E. J. (1982). Fading of immunofluorescence during microscopy: a study of the phenomenon and its remedy. *Journal of Immunological Methods*, **55**, 231–42.

Kabat, E. A. (1966). The nature of an antigenic determinant. *Journal of Immunology*, **97**, 1–11.

Kilmartin, J. V., Wright, B. & Milstein, C. (1982). Rat monoclonal antitubulin antibodies derived by using a new non-secreting rat cell line. *Journal of Cell Biology*, **93**, 576–82.

Knox, R. B. & Clarke, A. E. (1978). Localization of proteins and glycoproteins by binding to labelled antibodies and lectins. In *Electron Microscopy and Cytochemistry of Plant Cells*, ed. J.L. Hall, pp. 149–85. Amsterdam: Elsevier.

Kurn, N. (1981). Altered development by the multicellular alga *Volvox carteri* caused by lectin binding. *Cell Biology International Reports*, **5**, 867–75.

Lloyd, C. W. (1983). Helical microtubule arrays in onion root hairs. *Nature*, **305**, 311–3.

Lloyd, C. W. (1984). Toward a dynamic helical model for the influence of microtubules on wall patterns in plants. *International Reviews in Cytology*, **86**, 1–51.

Lloyd, C. W. (1986). Microtubules and the cellular morphogenesis of plants. In *Developmental Biology: A Comprehensive Synthesis Vol 2*, ed. L. Browder, pp. 31–58. New York: Plenum Press.

Lloyd, C. W., Clayton, L., Dawson, P. J., Doonan, J. H., Hulme, J. S., Roberts, I. & Wells, B. (1985). The cytoskeleton underlying side walls and cross walls in plants: molecules and macromolecular assemblies. *Journal of Cell Science*, Supplement 2, 143–55.

Lloyd, C. W. & Seagull, R. W. (1985). A new spring for plant cell biology: microtubules as dynamic helices. In *Trends in Biochemical Sciences*, 10, 476–8.

Mau, S. L., Raff, J. & Clarke, A. E. (1982). Isolation and partial characterization of components of *Prunus avium* L. styles including an antigenic glycoprotein associated with a self-incompatibility genotype. *Planta*, **156**, 505–16.

Millikin, B. E. & Weiss, R. L. (1984). Distribution of ConA binding carbohydrates during mating in *Chlamydomonas*. *Journal of Cell Science*, **66**, 223–39.

Murray, B. A., Ninan, H. L. & Loomis, W. F. (1983). Monoclonal antibody recognizing gp80 – a member glycoprotein implicated in intercellular adhesion of *Dictyostelium discoideum*. *Molecular and Cell Biology*, **3**, 863–70.

O'Brien, T. P. & McCully, M. E. (1981). *The Study of Plant Structure Principles and Selected Methods*. Melbourne: Termarcarphi Pty. Ltd.

Raff, J. W. & Clarke, A. E. (1981). Tissue specific antigens secreted by suspension cultured callus cells of *Prunus avium* L. *Planta*, **153**, 115–24.

Raikhel, N. V., Mishkind, M. & Palevitz, B. A. (1984). Immunocytochemistry in plants with colloidal gold conjugates. *Protoplasma*, **121**, 25–33.

Rider, C. L., Nuss, D. & Bowser, S. S. (1984). Correlative immunofluorescence and electron microscopy on the same section of epon-embedded material. *Journal of Cell Biology*, 99, 196a, abstract 728.

Roberts, E., Seagull, R. W., Haigler, C. H. & Brown, R. M. (1982). Alteration of cellulose microfibril formation in eukaryotic cells: Calcofluor White interferes with microfibril assembly and orientation in *Oocystis apiculata*. *Protoplasma*, **113**, 1–9.

Roberts, I. N., Lloyd, C. W. & Roberts, K. (1985). Ethylene-induced microtubule reorientations: mediation by helical arrays. *Planta*, **164**, 439–47.

Roberts, K., Burgess, J. B., Roberts, I. & Linstead, P. (1985*a*). Microtubule rearrangements during cell growth and development. In *Botanical Microscopy 1985*, ed. A. W. Robards, pp. 263–283. Oxford: Oxford University Press.

Roberts, K., Grief, C., Hills, G. J. & Shaw, P. (1985*b*). Cell wall glycoproteins: structure and function. *Journal of Cell Science*, Supplement 2, 105–27.

Roberts, K., Lloyd, C. W., Shaw, P., Johnston, A. W. B. & Woolhouse, H. W., eds. (1985*c*). *The Cell Surface in Plant Growth and Development*. *Journal of Cell Science*, Supplement 2.

Roberts, K., Phillips, J., Shaw, P., Grief, C. & Smith, E. (1985*d*). An immunological approach to the plant cell wall. In *The Plant Cell Wall*, eds. C. T. Brett & J. R. Hillman, Society for Experimental Biology Seminar Series No. 28, pp. 125–54. Cambridge: Cambridge University Press.

Robertson, J. G., Wells, B., Bisseling, T., Farnden, K. J. F. & Johnstone, A. W. B. (1984). Immunogold localization of leghaemaglobin in the plant cytoplasm in nitrogen-fixing root nodules of pea. *Nature*, **311**, 254–6.

Robertson, J. G., Wells, B., Brewin, N. J. & Williams, M. A. (1985). Immunogold localization of cellular constituents in the legume-rhizobium symbiosis. In *Oxford Surveys of Plant Molecular and Cell Biology*, vol. 2, ed. B. J. Miflin, pp. 69–89. Oxford: Oxford University Press.

Roth, J. (1983). The colloidal gold marker system for light and electron microscopic cytochemistry. *Immunocytochemistry*, **2**, 217–84.

Roy, A. (1984). Structural studies on the polysaccharide from *Streptococcus pneumoniae* type 20. *Carbohydrate Research*, **128**, C11–12.

Sengbusch, P. V., Mix, M., Wachholz, I. & Manshard, E. (1982). FITC-labelled lectins and Calcofluor White ST as probes for the investigation of the molecular architecture of cell surfaces. Studies on conjugatophycean species. *Protoplasma*, **111**, 38–52.

Sengbusch, P. V. & Muller, U. (1983). Distribution of glycoconjugates at algal cell surfaces as monitored by FITC-conjugated lectins. Studies on selected species from Cyanophyta, Pyrrophyta, Raphidophyta, Euglenophyta, Chromophyta and Chlorophyta. *Protoplasma*, **114**, 103–13.

Shaw, P. J. & Henwood, J. A. (1985). Immuno-gold localization of cytochrome f, light-harvesting complex, ATP synthase and ribulose 1,5-bisphosphate carboxylase/oxygenase. *Planta*, in press.

Simmonds, D. H., Seagull, R. W. & Setterfield, G. (1985). Evaluation of techniques for immunofluorescent staining of microtubules in cultured plant cells. *Journal of Histochemistry and Cytochemistry*, **33**, 345–52.

Smith, E., Roberts, K., Hutchings, A. & Galfrè, G. (1984*a*). Monoclonal antibodies to the major structural glycoprotein of the *Chlamydomonas* cell wall. *Planta*, **161**, 330–8.

Smith, E., Roberts, K., Butcher, G. W. & Galfrè, G. (1984*b*). Monoclonal antibody screening: two methods using antigens immobilized on cellulose. *Analytical Biochemistry*, **138**, 119–24.

Stone, B. A., Evans, N. A., Bonig, I. & Clarke, A. E. (1984). The application of Sirofluor, a chemically defined fluorochrome from aniline blue for the histochemical detection of callose. *Protoplasma*, **122**, 191–5.

Sutcliffe, J. G., Shinnick, T. M., Green, N. & Lerner, R. A. (1983). Antibodies that react with predetermined sites on proteins. *Science*, **219**, 660–6.

Tainer, J. A., Getzoff, E. D., Alexander, H., Houghton, R. A., Olson, A. J., Lerner,

R. A. & Hendrickson, W. (1984). The reactivity of anti-peptide antibodies is a function of the atomic mobility of sites in a protein. *Nature*, **312**, 127–34.

Tiwari, S. C., Wick, S. M., Williamson, R. E. & Gunning, B. E. S. (1984). Cytoskeleton and integration of cellular function in cells of higher plants. *Journal of Cell Biology*, **99**, 63s–9s.

Townsend, R. & Plaskitt, K. A. (1985). Immunogold localization of P55-fibril protein and P25 spiralin in *Spiroplasma* cells. *Journal of General Microbiology*, **131**, 983–92.

Vian, B., Brillouet, J. M. & Satiat-Jeunemaitre, B. (1983). Ultrastructural visualization of xylans in cell walls of hardwood by means of xylanase-gold complex. *Biologie Cellulaire*, **49**, 179–82.

Vreeland, V. & Laetsch, W. M. (1984). Intracellular synthesis of *Fucus* cell wall antigens. *Plant Physiology*, **75**, 62a.

Waaland, S. D. & Waaland, J. R. (1975). Analysis of cell elongation in red algae by fluorescent labelling. *Planta*, **126**, 127–38.

Wells, B. (1985). Low temperature box and tissue handling device for embedding biological tissue for immunostaining in electron microscopy. *Micron and Microscopica Acta*, **16**, 49–53.

Wetherbee, R., Hinch, J. M., Bonig, I. & Clarke, A. E. (1985). Response of *Zea mays* roots to infection with *Phytophthora cinnamomi*. II. The cortex and stele. *Protoplasma*, **126**, 188–97.

Wick, S. M. & Duniec, J. (1984). Immunofluorescence microscopy of tubulin and microtubule arrays in plant cells. II. Transition between the preprophase band and the mitotic spindle. *Protoplasma*, **122**, 45–55.

Wick, S. M., Muto, S. & Duniec, J. (1985). Double immunofluorescence labelling of calmodulin and tubulin in dividing plant cells. *Protoplasma*, **126**, 198–206.

Wick, S. M., Seagull, R. W., Osborn, M., Weber, K. & Gunning, B. E. S. (1981). Immunofluorescence microscopy of organized microtubule arrays in structurally stabilized meristematic plant cells. *Journal of Cell Biology*, **89**, 685–90.

Woodward, M. P., Young, W. W. & Bloodgood, R. A. (1985). Detection of monoclonal antibodies specific for carbohydrate epitopes using periodate oxidation. *Journal of Immunological Methods*, **78**, 143–53.

Young, W. W., Johnson, H. S., Tamura, Y., Karlsson, K–A., Larson, G., Parker, J. M. R., Khare, D., Spohr, U., Baker, D. A., Hindsgaus, O. & Lemieux, R. U. (1983). Characterization of monoclonal antibodies specific for the Lewis, a human blood group, determinant. *Journal of Biological Chemistry*, **258**, 4890–4.

J.H. DOONAN and L. CLAYTON

Immunofluorescent studies on the plant cytoskeleton

Introduction

The cytoskeleton is a complex network of protein filaments that structures and organizes the cytoplasm of eukaryotes (Stossel, 1984; Steinart, Jones and Goldman, 1984; Olmstead *et al.*, 1984; Dustin, 1984). At least three distinct cytoskeletal systems are recognised in animal cells: microtubules (MT), actin microfilaments and intermediate filaments. Plant cells also contain MTs, microfilaments and possibly an intermediate filament-related system. Cytoskeletal proteins have been remarkably conserved throughout evolution, both in antigenicity and in terms of the supramolecular structures they form within the cell, which reflects the central role they play in cell division and cell shaping.

Antibodies to these cytoskeletal proteins have been used to investigate cytoskeletal organisations using immunofluorescent localization techniques and have made significant contributions to our understanding of the cytoplasmic structuring of animal cells (Aubin, 1981). The advantage of immunofluorescence techniques over conventional ultrastructure studies lies in their enabling the examination of specific components within whole cells and tissues in three dimensions – a feature which is very useful when considering plant cells.

Immunofluorescent techniques exploit the ability of antibodies to recognise and bind to specific antigens. Antibodies can be raised against cytoskeletal proteins and applied to suitably treated cells where they bind to structures containing those antigens. To visualize the location of the antigens one can either use antibodies directly linked to a fluorochrome (direct immunofluorescence) or a fluorochrome conjugated to an antibody which recognises the primary antibody (indirect immunofluorescence, IIF). The latter method is generally preferred since there is amplification of the signal. Details of how these techniques are applied to the location of various antigens within plant cells can be found in a recent review by Knox (1982) (see also the chapter by Perrot-Rechenmann and Gadal, this volume).

The techniques and practical aspects of immunofluorescence staining of the cytoskeleton as applied to animal cells are comprehensively described by Osborn & Weber (1982). It is only relatively recently, however, that such methods have been applied to the study of plant cells; the reason being that the preparation of plant cells has posed a number of technical problems not encountered when using animal tissues. Most plant cells are surrounded by a cell wall which presents a barrier to the entry of antibodies; consequently the initial immunofluorescence studies on plant cells are performed on wall-less endosperm cells (Franke *et al.*, 1977), algal gametes (Weber *et al.*, 1977) and protoplasts (Marchant, 1978; Powell *et al.*, 1980; Van der Valk *et al.*, 1980). Subsequently, Wick *et al.* (1981) devised fixation and permeabilization strategies which have enabled the immunofluorescent staining of microtubules in intact cells from a complex tissue, i.e. root tip meristems (Wick *et al.*, 1981, Wick & Duniec, 1983). Minor modifications of this protocol have permitted staining of MTs in a number of plant cell types, e.g. root hairs (Lloyd, 1983; Lloyd & Wells, 1985), pea epidermal cells (Roberts, Lloyd & Roberts, 1985), pollen mother cells (Dickinson & Sheldon, 1984), tissue culture cells (Simmonds, Setterfield & Brown, 1983), moss filaments (Doonan, Cove & Lloyd, 1985) and fern cells (Doonan, Duckett & Lloyd, 1986).

Most of these immunofluorescent studies of the cytoskeleton have concentrated upon the arrangement of the microtubules and antibodies used have generally been raised against tubulin from non-plant sources, usually from mammalian brain. Despite a high degree of evolutionary conservation, however, antisera against tubulin from animal sources are not guaranteed to cross-react with plant microtubules (Lloyd & Doonan unpublished observations). The monoclonal antibodies (McAb), YOL 1/34 and YL 1/2, raised against yeast tubulin (Kilmartin, Wright & Milstein, 1982) recognise tubulin from a wide range of organisms and are now widely used for routine localization of tubulin and microtubules in plant and animal cells and microbes.

Unfortunately, most of the above techniques involve squashing the tissues and so disrupting and destroying the spatial relationships between individual cells within the tissue. With less complex tissues e.g. moss filaments, one may immobilise the entire plant on a glass support and so retain continuity between cells throughout the procedure (Doonan *et al.*, 1985). Complex multicellular organs may be embedded in PEG and thick sections stained (Hawes & Horne, 1985) which maintains cell-to-cell continuity but sacrifies some advantages of the whole cell images provided by the method of Wick *et al.* (1981).

It is against the wider background of development that individual cellular events become meaningful. The elucidation of cell patterning by such techniques, therefore, is important to plant morphogenesis. The formative regions of most plants are composed of ordered files of cells which arise by precisely co-ordinated cell division and cell elongation. In such morphogenetic processes microtubules appear to play an important role. MTs are a major component of the spindle and the phragmoplast – the MT assemblies which partition the nucleus and cytoplasm. MTs are closely associated also with the plasma membrane and are believed to influence the orientation of the cellulose microfibrils within the wall. Thus they control the mechanical properties of the wall and overall cell expansion (see Robinson & Quader, 1982).

This chapter will consider the contributions which IIF studies have made to our understanding of cytoplasmic structure and organisation, particularly the microtubules, and how these are related to morphogenetic processes such as cell elongation and cell division.

Cell elongation

The growth of plant tissues is, in part, the result of cell expansion within the tissue. The force for cell expansion is the hydrostatic or turgor pressure generated by the solutes within the cell and, to prevent cell lysis, this pressure is contained by the cell wall. The hydrostatic pressure acts uniformly on all faces of a cell. Hence, by locally modifying the mechanical properties of the wall, the cell may achieve directed expansion (i.e. elongation).

There are at least two modes of cell elongation found in plants; cells within multicellular tissues or derived from such tissues tend to increase in length more or less uniformly along their axis of elongation, whereas many filamentous cells increase in length only at or close to their apex. "Uniform" cell elongation is found mainly in meristems of higher plants, ferns and bryophytes as well as in many tissues derived from meristems, although some filamentous algae appear to have a similar mode of growth. Tip elongation is displayed by some specialised higher plant tissues – root hairs and pollen tubes – and many filamentous tissues of lower plants (i.e. moss protonemata and fungal hyphae). The whole cell view afforded by IIF allows the direct comparison of MT organisation in cells expanding by different modes (Fig. 1). Uniformly elongating cells tend to have fairly regular cortical MT arrays, usually transverse to the axis of elongation, while the tip expanding cells have more axial arrays which form foci within the apical dome.

Since Ledbetter & Porter's original report (1963), in which plant MTs were first convincingly seen by electron microscopy (EM), examinations of numerous other cell types have confirmed that the phenomenon of cortical

MTs paralleling the innermost layer of cellulose microfibrils is widespread (Robinson & Quader, 1982). The removal of MTs by colchicine leads to cellulose microfilaments subsequently being laid down in a random manner – no longer transversely – and concomitantly, the cells become spherical (Pickett-Heaps, 1967). Cellulose continues to be deposited but in the absence of MTs is no longer effective in controlling the orientation of cell expansion. The coincidental orientations of MTs and cellulose microfibrils and the apparent requirement of MTs for orientated cellulose deposition have encouraged the notion that MTs spatially co-ordinate cell wall synthesis and hence determine the expansion characteristics of the cell. The removal of MTs by colchicine (Pickett-Heaps, 1967) causes spherical rather than elongated cell shape. To achieve effective co-ordination, the MTs must be integrated on a cellular level, responding to endogenous and exogenous signals as a unit. Thus the orientation of the cellulose microfibrils relies on the MTs acting in a co-ordinated fashion.

Fig. 1. The microtubule arrays of two plants displaying differing patterns of cell expansion: compared by staining with monoclonal anti-tubulin antibodies (a) onion root cell containing MTs which run transversely around the cell cortex – at right angles to the axis of cell elongation. Through focusing on a moss protonemata (b) demonstrates a complex meshwork of MTs throughout the depth of the cytoplasm, but whose orientation is approximately along the axis of elongation, except at the apical dome where the MTs come together forming foci. Bar = 10 μm.

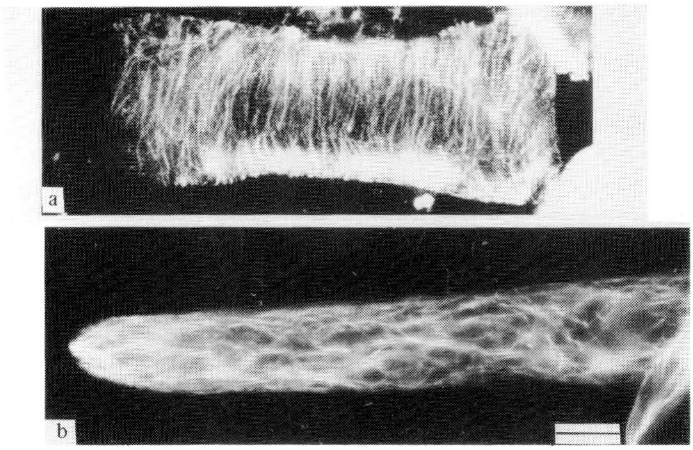

Cortical MTs and "uniform" cell elongation

The interphase MTs of uniformly expanding cells form cortical arrays closely adpressed to the cell membrane. Very few MTs are found deep within the cytoplasm of these cells during interphase. That the interphase cortical MT array (ICMA) was an integral assembly and not composed of unconnected parts was demonstrated using an antiserum to brain tubulin on carrot suspension culture cells which had been briefly treated with cellulases (Lloyd *et al.*, 1980). MTs in elongated cells appear to form transverse hoops around the short axis of the cell but, as such unfixed cells became spherical during treatment with cellulase, the MTs were observed to maintain contact with the plasma membrane and with each other in a coherent assembly without apparent free ends. Hence the ICMA behaves as a unit and can certainly accommodate drastic changes in cell shape. This behaviour is consistent with the MT array being based on helices rather than hoops or disconnected MTs and demonstrates the advantages of whole cell immunofluorescent techniques for following complex systems such as the microtubule cytoskeleton. Through-focusing allows one to build up a complete 3-D layout of the system quickly whereas the same information could not be obtained unambiguously from thin sections examined by EM.

EM studies often find that MTs may occupy oblique and longitudinal orientations (Newcomb & Bonnett, 1965; Seagull & Heath, 1980). Comparing the ICMAs of several cell types has led Lloyd (1984) to suggest that ICMAs of most, if not all, higher plant cells can be thought of as helical. The transverse arrays of carrot cells (Lloyd *et al.*, 1980) or onion meristem cells (Wick *et al.*, 1981) could be interpreted as a helix of shallow pitch and the net axial or longitudinal MTs of a radish root hair (Newcomb & Bonnett, 1965) as a steeply pitched helix. The 45 ° multistart helix seen in onion root hairs is intermediate between these two extremes of helical winding and establishes the multi-start helix as a suitable model for the ICMA of uniformly expanding cells (Lloyd, 1983). The helical model of MT organisation can account for most, if not all, of the orientations of cortical MT simply by winding or unwinding (i.e. changing the pitch of the helix). Ethylene causes such a change – from transverse arrays to longitudinal arrays – in pea epidermal cells. This is concomitant with a similar switch in cellulose microfibril orientation and the cells expand laterally (Steen & Chadwick, 1981) – no longer being constrained by transverse inner wall microfibrils. Sheets of untreated epidermal cells stained with the anti-tubulin antibody, YOL 1/34, demonstrated the presence of transverse arrays within nearly all cells (Roberts, Lloyd & Roberts, 1985). Within 30 minutes of ethylene application cells containing oblique MT arrays (similar to those observed in onion root hairs; Lloyd, 1983) are found in increased numbers and axial MT arrays are also observed. The whole cell view of these MT arrays demonstrates the helical nature of the intermediate 'oblique' arrays and that the ICMA responds as a unit with no cell containing both a transverse and a longitudinal array. Since no cell is seen without MTs during this transition, the idea that the ICMA unwinds rather than changing its configuration by de- and then re-assembly of individual MTs is reinforced.

The implications of helical MT arrays for the structure and mechanical properties of walls and for cell expansion have recently been extensively discussed (Lloyd, 1984). Many problems remain to be solved before we understand the complex relationship between cortical MTs and cell expansion, i.e. the molecular mechanisms which determine and alter the pitch of the MT helices, the interaction between the MTs and the cell wall and the arrangement of MTs relative to one another (MT-MT distance). The spacing between MTs within the ICMA is particularly interesting as it impinges on the process of xylogenesis and on pre-prophase band (PPB) formation (see chapter by Roberts, this volume). During differentiation of xylem elements in *Zinnia* (Falconer & Seagull, 1985) the spacing of MT bundles within the ICMA is modified such that the MTs associate laterally to form bands or spirals around the cell. These patterns of MTs predict sites of wall thickening. The significance of MT-MT spacing in the ICMA is unclear and requires further investigation.

Tip-growing cells

The larger parts of cell expansion occurs at or near the tip of many filamentous plant cells. The tip of such cells is a hemispherical dome which constantly expands in such a way as to increase cell length and maintain a constant cell diameter. The association of MTs with the apical dome of such cells is poorly documented. EM studies have generally emphasised the lack of cytoskeletal elements in the apical dome of moss (Schmiedel & Schnepf, 1980) and that of root hairs (Newcomb & Bonnet, 1965; Seagull & Heath, 1980), or have stressed the lack of any obvious organization of MTs near the dome (DeMaggio & Stetler, 1977). This may be due to poor fixation conditions used for the electron microscopy since IIF (using the antibody, YOL 1/34) showed that the apical dome of radish root hairs is especially sensitive to sub-optimal fixation; under suitable conditions MTs are present at the very tip (Lloyd & Wells, 1985). In the tip cells of protonemal filaments of the moss *Physcomitrella*, the MTs run in an axial direction, undulating throughout the cytoplasm (Fig. 1*b*). The MTs from a three-dimensional meshwork which envelops the nucleus and plastids extend into the apical dome. Within the dome, the microtubules converge into one, or a few, foci at or very close to, the surface as indicated by IIF staining with the antibody, YL 1/2. Root hairs contain MT arrays within the apical dome which also form foci (Lloyd, 1983; Lloyd & Wells, 1985). The cortical MTs form a steeply-pitched trellis work but this is usually less regular than in uniformly elongating plant cells.

The maintenance of a uniform diameter in protonemata requires a normal MT array within the apical dome. The carbamate herbicide, CIPC, which disrupts microtubule organising centres (MTOCs) in evolutionary diverse phyla (Coss & Pickett-Heaps, 1974; Oliver, Krawiec & Berlin, 1978) induces abnormal MT arrays in the tip and leads to deformed cell shape (Fig. 2). The response varies with the concentration of drug – low

Fig. 2. The MTOC-disrupting drug, CIPC at 1 μM, causes aberrant foci of MTs within the apical dome of *Physcomitrella* protonemata (a) and (b). Although other cytoplasmic microtubules are diminished in numbers and staining intensity (a) the spindle (marked with an asterisk) of such cells appears to be normal (b). Effects of CIPC on cell shape include bending and deformation of the tip (a) or at higher concentrations loss of uniform cylindrical shape (c). Higher concentrations (c) led also to abnormal MT arrays throughout the cell. Bar = 10 μm. (Fig. 2 adapted with permission from the *Journal of Cell Science*, **75**, in press.)

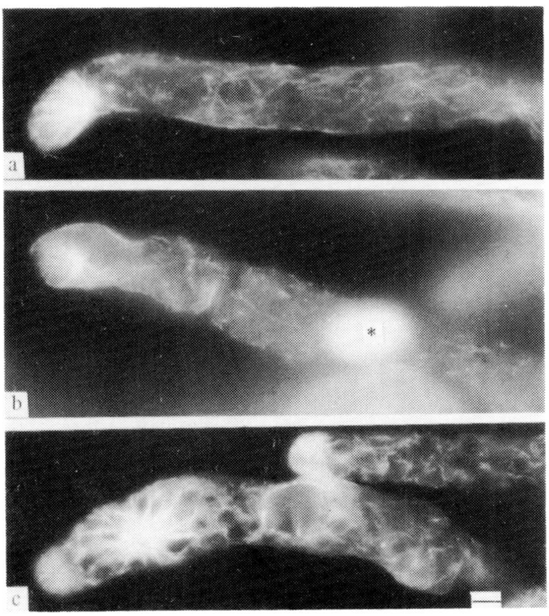

concentrations (1 μM) lead to increased numbers of MTs at the tip and cause bending (Figs. 2*a* & 2*b*). Higher levels cause aberrant foci of MTs ("starbursts") and are correlated with a loss of uniformity of cylindrical shape of the cell. Prolonged culture in this drug induces multiple abortive tips. Thus MTs appear to be organised at the tip of protonemata, but whether they originate or terminate there is not known. In contrast to the interphase MT arrays of protonemata, the ICMA of uniformly elongating cells is at least two orders of magnitude more resistant to CIPC (Clayton & Lloyd, 1984) further emphasizing the different organization of MT arrays in cells which expand by alternative modes.

Unlike the uniformly expanding cells of higher plants, moss protonemata maintain MT arrays within their tips throughout the cell cycle, although these may be fewer in number during mitosis. Even during mitosis, these tip MTs remain focused and sensitive to the disorganizing effects of CIPC which suggests they are functional in maintaining cell shape throughout the cell cycle. The ICMA of uniformly elongating cells disappears entirely in mitosis and is re-established late in cytokinesis. The maintenance of cytoplasmic MT arrays in mitotic protonemata presumably reflects the requirement of free-living cells to maintain their cell morphology without the mechanical support of neighbouring cells. Cells within a tissue may depend (at least partially) on their neighbours to prevent undirected cell expansion during mitosis.

The control of cell shape is important throughout the cell cycle if ordered files of cells are to develop. The different growth strategies, tip growth and "uniform" cell elongation, both achieve ordered files by differing use of what are basically the same components in both cells – microtubules and cell wall.

Cell division

Cell division is usually strictly co-ordinated with cell expansion during development, both in terms of timing and positioning. The orientation of the division plane determines the initial spatial relationship between the two daughter cells and thus represents one possible means of influencing development. MTs are intimately involved in the implementation of cell division, undergoing an orderly and predictable series of changes in organization, firstly forming the spindle which separates the chromosomes and then the phragmoplast which is involved in cell plate formation. Although the manner of spindle initiation and the nature of accessory MTs vary considerably amongst the different cell types, the spindle and phragmoplast are remarkably consistent in structure.

Fig. 3. MT arrays during the transition in onion meristem cells from interphase to mitosis. (a) – (c) probable stages in the formation of pre-prophase bands from interphase cortical MT arrays.

(a) Late interphase cortical MT array orientated around the short axis of the cell – note the lack of MTs near each end of the cell.

(b) Broad "PPB" – MTs at the edge of the PPB are still discrete and the nuclear envelope is only faintly fluorescent. MTs near the ends of the cell have decreased markedly in numbers.

(c) The PPB as an equatorial band of MTs around the cell, linked to the nucleus by many bundles of MTs. The nuclear envelope fluoresces brightly with anti-tubulin antibody and is connected by numerous fibres which run towards the cell cortex.

(d) Prophase spindle after the PPB has completely disappeared. Note the symmetrical poles and lack of cytoplasmic MTs.

(e) The metaphase spindle has broad poles and spindle fibres extend to equatorial plane. Unlike animal cells no centrioles are present at the poles of higher plant spindles. Bar = 10 μm.

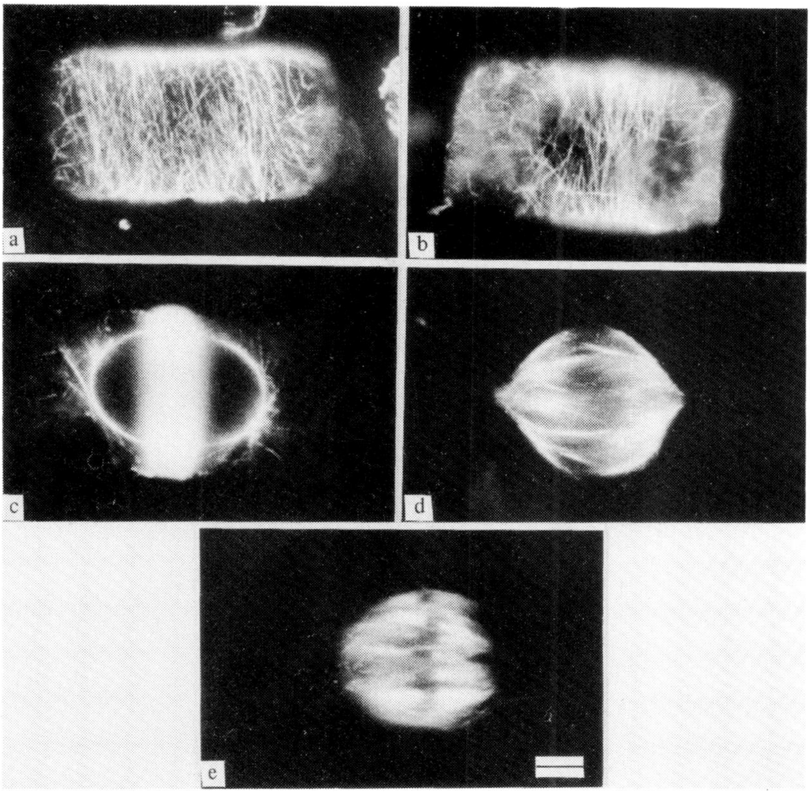

Spindle initiation

The transition from interphase to mitosis has been rather poorly characterized by EM, perhaps due to the difficulty of finding and recognising intermediate stages and interpreting the complex re-arrangements which occur during this period. There are now several IIF studies which provide details on this period in onion root cells (Wick and Duniec, 1983; 1984) and in endosperm cells (Schmit *et al.*, 1983).

The PPB under the electron microscope is an equatorial band of cortical MTs which forms at the site where the future cell plate will fuse with the parental wall (Pickett-Heaps & Northcote, 1966). In onion cells the cortical MTs (Fig. 1*a*) give way to the PPB band visualized using the McAb, YOL 1/34 (Fig. 3*b* and 3*c*). The PPB forms just prior to the symmetrical development of the two spindle poles of the prophase spindle (Fig. 3*d*). The poles are compact in prophase, widening during metaphase (Fig. 3*e*) as the spindle fibres extend towards the equator. The spindle fibres often associate laterally into sub-groups, thus the broad poles of the spindle may appear to be composed of several 'mini-poles'.

Physcomitrella protonemata, which have complex cytoplasmic MT arrays, do not form a PPB. Prior to mitosis in apical cells, a focus of MTs is formed to the anterior side of the nucleus and can be stained using the McAb, YL 1/2 (Fig. 4*a*). In sub-apical cells the nucleus migrates to the apical end of the cell prior to division and during this process the nucleus becomes associated with numerous bundles of MTs which link it to the division site (Fig. 4*b*). As the nucleus approaches its division site, the MTs also form a focus to its anterior side (as in the apical cell) and MTs run back across the nuclear surface (Doonan, Lloyd & Cove, unpublished data). The spindle seems to evolve from these MTs (Fig. 4*c*). The metaphase spindle has broad poles like those in onion cells, but is still associated with cytoplasmic MTs (Fig. 4*d*) which run from the poles to the cell extremities.

It is clear that protonemata and onion cells initiate their spindle arrays in different ways, perhaps as a result of the constraints placed upon them by their pre-existing interphase arrays. MT arrays just prior to spindle formation often appear to be re-arrangements of the interphase arrays (compare Fig. 3 and Fig. 4 with Fig. 1), often giving the appearance of concentrating the no longer necessary cytoplasmic MTs close to where the spindle will form.

Onion cells develop a PPB and their spindle poles develop symmetrically, whereas moss protonemata lack a PPB and their spindles appear to develop with an inherent asymmetry. Onion cells divide to produce two daughter cells each of which is generally capable of immediate expansion and division whilst only the apical daughter cell of a protonemal cell

Fig. 4. The probable transition from interphase to mitosis and cytokinesis in *Physcomitrella* protonemata, showing the changing association between the nucleus and MTs. The nucleus is marked by an asterisk.

(a) In apical cells MTs form a focus to the anterior of the nucleus and from this focus MTs spread back across the nuclear surface.

(b) Nuclear migration in sub-apical cells involves the formation of MT bundles between the nucleus and the side branch initiation site (=division site).

(c) The microtubules to the anterior of the nucleus, increase in numbers and begin to spread across the surface of the nucleus.

(d) The metaphase spindle is barrel-shaped and cytoplasmic MTs are connected to its broad poles.

(e) During anaphase the number of pole to cortex MTs (arrowed) increases dramatically concomitant with a swivelling of the spindle to an oblique orientation.

(f) The spindle-shaped early phragmoplast lays down an oblique cross-wall (between arrowheads), thus reflecting the re-orientation of the spindle which occurs during anaphase (e).

Throughout mitosis, microtubules are continually present within the apical dome (a), (b), (d), (e) and (f). Bar = 10 μm. (Figure 4 reproduced in part with permission from *Journal of Cell Science*, vol. 75).

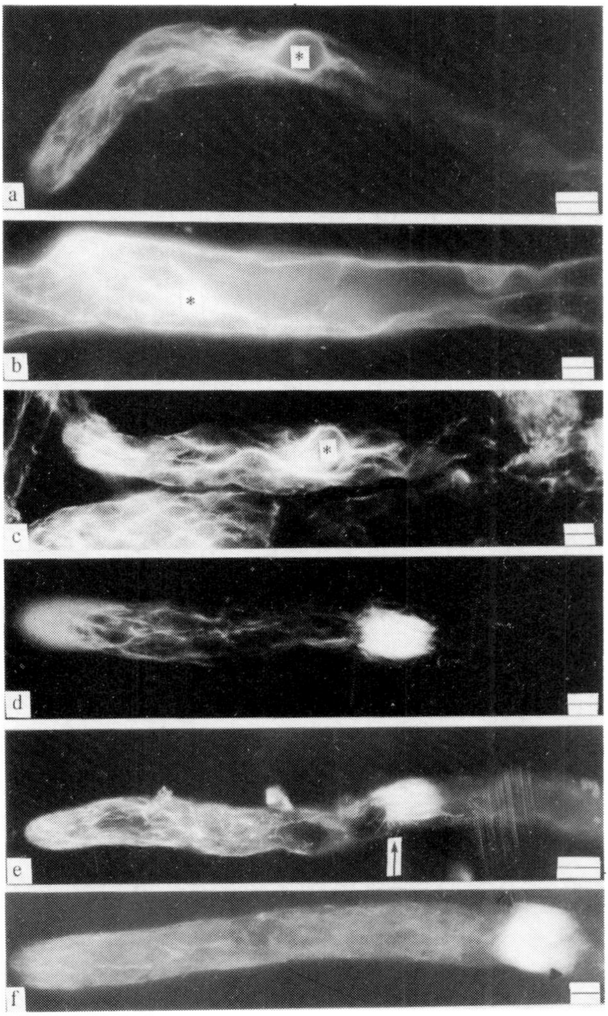

division immediately expands and divides. Perhaps the asymmetry of spindle formation in moss protonemata reflects the unequal division which subsequently occurs.

Position and plane of division

A fundamental problem of plant developmental biology is how and when the plane and position of division is determined. In meristematic plant cells, the PPB has been found to occupy the site where the cell plate fuses with the side walls in both symmetrical and asymmetrical divisions of *Azolla* roots (Gunning, Hardham & Hughes, 1978), as well as in numerous other species (recently reviewed by Gunning, 1982). Thus the PPB has been proposed as part of a mechanism for determining cell plane position with multicellular organs. The failure of EM techniques to locate PPBs in tissues growing in a disordered fashion (i.e. suspension culture cells, callus and dividing protoplasts) or in simple filamentous tissues such as moss protonemata led to the correlation of PPB with higher degrees of morphogenetic order (Gunning, 1982). However, using IIF techniques to screen relatively large numbers of cells, PPBs have been observed in suspensions of cultured cells of *Zinnia* (Falconer & Seagull, 1985) and of *Vicia* (Simmonds, Setterfield & Brown, 1983). Division features are rare in these materials since the cells spend long periods in interphase; many of the PPB are not quite as compact as observed within tissues and so would be unrecognised by EM observations.

So far, IIF studies using a variety of anti-tubulin antibodies have shown a very close correlation between an ICMA (or rather the lack of endoplasmic MTs during interphase) and the formation of PPB, prior to mitosis. The ICMA may collapse to the cell equator – possible intermediate stages of such are often observed (Fig. 3*a*–*c*). In studies on *Azolla* root meristems in which every cell may be identified due to the highly predictable sequence of cell divisions – PPBs were observed which appeared perpendicular to the orientation which the cortical MTs were known to have until that stage (Gunning, Hardham & Hughes, 1978). Thus it was considered that the PPB could not, at least in these cases, be formed by bunching up of the ICMA. However, given its dynamic nature, the ICMA could conceivably switch orientation through 90 ° just prior to PPB formation and then bunch up. The function of the PPB is unknown, but if nothing else, it appears to be a marker for one otherwise cryptic cytoplasmic polarization which eventually influences the division plane (Pickett-Heaps, 1969). It has long been recognised that the PPB may be the effect rather than the cause of cytoplasmic polarization. In line with this, although the PPB coincides with the cortical site where the cell plate and side walls normally fuse, the presence of a PPB does not appear necessary for fusion to occur. Treatment of cells with CIPC leaves the ICMA and the PPB unaffected, but causes tripolar spindles and tri-limbed phragmoplasts (Figs. *5c* and *d* which lay down cross-walls in orientations unpredicted by the PPB (Clayton & Lloyd, 1984). The "extra" cross-walls may fuse with walls which would not have had an underlying PPB. Thus the cell plate is capable of fusing with regions of the cell wall other than those which were beneath the PPB.

The cell plate is laid down in specific orientations in moss protonemata

Fig. 5. The effect of CIPC on the spindle and phragmoplast of onion cells and moss protonemata. 5μM CIPC eliminates the pole to cortex MTs from moss protonemata, causing disorientated spindles (a) leading to disorientated phragmoplasts – the phragmoplast shown in (b) is about 90° to the normal orientation in these cells. CIPC also causes multipolar spindles in onion cells, (c) leading to abnormal numbers of nuclei which are separated by branched phragmoplasts in onion (d) and moss protonemata (e).

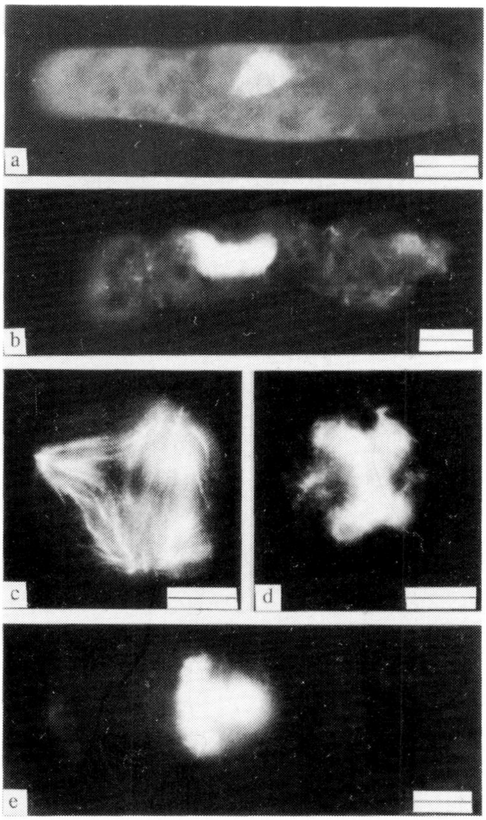

without the intervention of PPB and is sensitive to the effects of CIPC in a similar way to onion cells (Figs. 5*c* and *d*; Doonan *et al.*, 1985). When the apical cell divides, the spindle tilts during anaphase/telophase so that the phragmoplast, and hence the cell plate, is oblique to the filament axis (Jensen, 1981). The reorientation of the spindle is concomitant with an increase in the numbers of MTs running from the spindle pole to cell cortex (Fig. 4*e*) – the implication being that they are responsible for re-orientating the spindle. These pole to cortex MTs are very sensitive to CIPC and, if thus removed, the spindle becomes randomly orientated with respect to the cell axis (Fig. 5*a*). Phragmoplasts may attempt to divide the cell along its axis (Fig. 5*b*) as opposed to the normal transverse and oblique directions. Higher concentrations cause tri- and multi-polar spindles as in onion cells (Figs. 5*c*–*f*). However, in both cases, prior to division, the nuclei seem to be positioned relative to a site on the cell cortex. The spindle is retained in this position (in moss by the pole to cortex MTs) and, after daughter nuclei are reconstituted, the phragmoplast bisects them symmetrically. When cells are treated with high concentrations of CIPC which cause the formation of abnormal numbers of nuclei, the phragmoplasts formed in these cells appear to bisect the space between neighbouring nuclei regardless of whether the nuclei are misplaced or are smaller and more numerous than normal. Each stage in the process seems dependant on the one before being correctly positioned and mistakes (induced by drugs) are propagated through the entire division process.

The pole-cortex MTs of uniformly expanding cells disappear in late prophase. So, what maintains the spindle's position during the remainder of mitosis? Further investigation of the nature and role of nuclear-cortex cytoskeletal connections is required, the PPB possibly being irrelevant but nevertheless a convenient marker for the cortical site to which the nucleus and spindle is anchored during mitosis.

The most obvious approach is to use antibodies against other cytoskeletal systems and cytoskeleton-associated proteins to probe their possible involvement.

The PPB – its relationship to other MT arrays

Amongst land plants at least, the possession of a PPB by a cell is very closely correlated with the possession of regular interphase cortical MT arrays – IIF shows very few, if any, endoplasmic MTs in these cell types during interphase. Cells which have abundant endoplasmic MTs (even though they may have cortical MTs too) seem to lack PPBs. The cortical MTs often seem less regular, perhaps integrated with the endoplasmic arrays as may be the case in moss protonemata.

An example of a higher plant cell which lacks PPBs is endosperm cells (Schmit *et al.*, 1983). Staining with anti-tubulin antibodies reveals the lack of regular ICMAs and, instead, the presence of complex endoplasmic MTs during interphase.

During the development of *Physcomitrella* (Doonan *et al.*, 1985, Doonan, unpublished data) and possibly other mosses (Schnepf, 1973; Schnepf cited by Gunning, 1982), PPBs are found in the leaves and shoots but not in the protonemata. Leafy shoots form from side branch initials of protonemata which have been treated with cytokinin. Anti-tubulin antibody staining of the young leafy shoots reveals MT arrays (Fig. 6*a* and *b*) which lack the foci near the apex characteristic of protonemata, but which still retain endoplasmic MTs and so lack PPB. Cell squashes of yet older buds contain cells which have MT arrays typical of higher plant cells (i.e. ICMAs, PPBs and spindles without associated cytoplasmic MTs, Fig. 6*c*). The mature leaves of the moss have transverse ICMAs but in the apical cell of each leaf the MTs form a multi-start helical array (Fig. 6*d*). There is no evidence, either from EM or IIF, that endoplasmic MTs are common in these cells during interphase.

Similarly in ferns, PPBs and cortical MTs seem to co-distribute in the early stages of the gametophyte (Wada *et al.*, 1980) but during spermatogenesis, cells develop endoplasmic as well as cortical MTs and so the ability to form a PPB is lost (Doonan *et al.*, 1980). Thus cells appear to be able to form a PPB only if they have regular cortical MT arrays, possibly by some sort of bunching up.

Implications for MT organization

The PPB and the ICMA have several properties in common – their organization is similar, both being composed of cortical, more or less co-aligned MTs and both are highly resistant to CIPC. Evidence that both the PPB and the ICMA have a different basis for their organization from that of other MT arrays has recently been shown using an antibody which recognises the MT organizing material of animal cells (Calarco-Gillam *et al.*, 1983). This antibody, 5051, locates putative MT-nucleating material at the spindle poles of onion cells throughout mitosis and also on the perinuclear region during and just after cytokinesis (Clayton, Black & Lloyd, 1985). No staining is associated with either the PPB or the mature ICMA, implying that no new MTs are formed either during re-orientations of cortical arrays or during PPB formation (although the possibility that individual MTs may lengthen or shorten is not precluded).

The MTs of the ICMA appear to be formed late in cytokinesis, nucleated close to the surface of the daughter nuclei as the phragmoplast reaches the

Fig. 6. Changes in MT arrays during the development of shoots from *Physcomitrella* protonemata. (a) Side branch initial containing a meshwork of endoplasmic and cortical MTs. Its developmental fate is undetermined and could produce either filamentous or shoot tissues. (b) Multicellular bud containing an interphase cell (marked by asterisk) which has both endoplasmic and cortical MTs, a mitotic cell (arrowed) in which there is a spindle and some (but fewer than in protonemata) cytoplasmic MTs, and a cell (arrowhead) in the process of division by means of a phragmoplast. (c)–(e) Cell squash from older multicellular buds (c) interphase cell containing solely cortical MTs, (d) a PPB-like array and (e) mitotic cells containing a spindle without any associated cytoplasmic MTs (arrowed) and a phragmoplast (arrowhead). (f) A moss leaf stained for MTs – only cortical MTs are observed. In various cells within the leaf, the MTs may be arranged differently, reflecting the diversity of organization found in higher plant cells. The MTs with the tip cell (arrowed) are oblique to the cell's long axis, while a favourable view of an internal leaf cell (arrowhead) shows its MTs to be largely transverse. Bar = 10 μM.

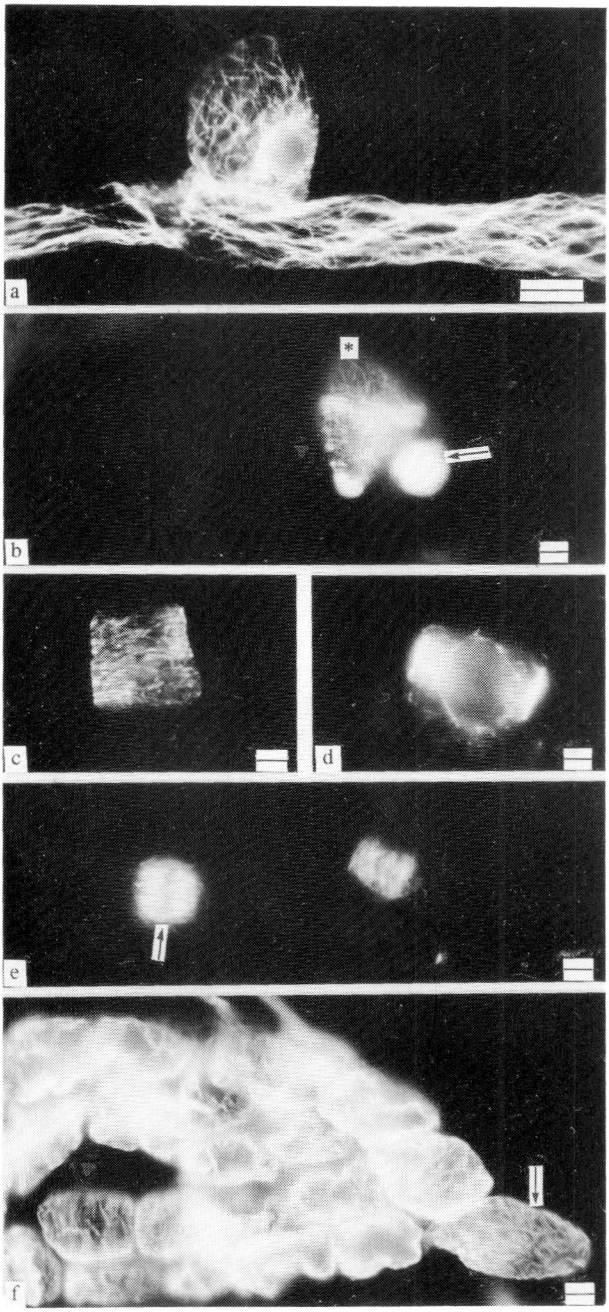
a
b
c
d
e
f

cell edge (Fig. 7*c*). 5051-positive material coats the nucleus at this stage (Clayton *et al.*, 1985). MTs can be seen radiating out from the nuclear surface in moss protonemata (Fig. 7*d*) and from the nucleus of developing pollen grains (Dickinson & Sheldon, 1984) and endosperm (Schmit *et al.*, 1983), so this may be a general means of assembling cytoplasmic MTs after

Fig. 7. Cytokinesis in onion (a)–(c) and in moss (d) involves the separation of the daughter nuclei (marked by an asterisk) and the partition of the cytoplasm by the phragmoplast. The phragmoplast is initiated in the mid-zone of the spindle between the two nuclei and is brightly stained with anti-tubulin antibody (a). The cell plate does not stain. The phragmoplast spreads outwards eventually meeting the cell periphery where remnants (arrowed) can be seen by anti-tubulin staining (b) as the interphase array is being reformed. Cytoplasmic MTs appear to arise from the nuclear surface and extend outwards towards the cell cortex (c). At this stage the entire nuclear surface stains with the antiserum, 5051, (c(i)); the nuclei of the same cell having been stained with DAPI (c(ii)).

Even though many cytoplasmic microtubules remain throughout mitosis in moss protonemata, additional microtubules appear to arise from the nuclear surface and radiate out into the cytoplasm (d). Bar = 10 μm.

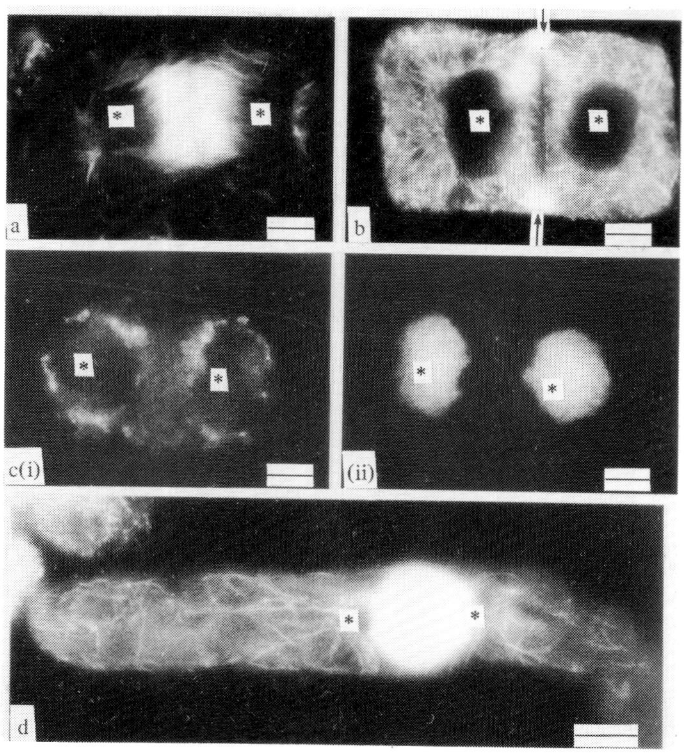

mitosis. In onion cells, however, the MTs nucleated in this way are subsequently organized around the cortex to produce the ICMA but little is known about the molecular basis of the interaction between the MTs and the cell cortex.

The phragmoplast is initiated in the mid-zone of the spindle (Fig. 7*a*), between the daughter nuclei, prior to the nucleation of cytoplasmic MTs. At this stage, the 5051-positive material appears only on the opposing faces of the daughter nuclei. As the phragmoplast spreads outwards, the 5051-positive material remains on the nuclei, spreading over their entire surface (see Fig. 7*b*). Only the early phragmoplast is associated with the 5051-positive material. The late phragmoplast shares with the ICMA and PPB the lack of any 5051-positive material and, incidentally, high resistance to disruption by CIPC.

The phragmoplast consists of many short co-aligned MTs. Both the other arrays not associated with 5051 staining are also, to some degree, co-aligned and show control over the spacing between their constituent MTs. These MTs are probably stabilized, possibly by cross-linking either to each other or other cytoskeletal elements and may be independent of new MT nucleation. The phragmoplast contains both actin (Clayton & Lloyd, 1985) and intermediate filament-related antigens (Dawson, Hulme & Lloyd, 1985) which may help stabilize the structure.

Perhaps it is not surprising that the organisation of plant MTs displays some novel features which do not seem to be present in wall-less cells. The three MT arrays for which MT nucleation appears to be spatially and temporally distinct from subsequent organization – the phragmoplast, ICMA and PPB are all closely associated with cell wall modification. This is a process which animal cells lack and has required a novel solution in terms of cytoskeletal organization in plants.

Prospects

IIF has greatly improved our understanding of the plant microtubule cytoskeleton and although in some cases the data may already be known from other sources such as EM, the ability to view whole cells has unravelled many of the knots that are created when trying to reconstruct the outlay of complex fibrous meshworks from thin section micrographs.

So far, most investigations have concentrated on the microtubule cytoskeleton and associated antigens. However, as antisera which recognise the other components of the plant cytoskeleton become available, they should open up new fields of investigation. Already some initial investigations have taken place on an intermediate filament-like system in plant cells (Dawson *et al.*, 1985). They used a monoclonal which recognises all classes of intermediate filament in animal cells (Pruss *et al.*, 1981) and demonstrated a close correlation in the distribution of this antigen to that of microtubules. The distribution of actin filaments has also been studied in plant cells but usually these studies have relied on fluorescent derivatives of the fungal toxin phallodin although Clayton & Lloyd (1985) have confirmed the results obtained from the toxin with monospecific antibodies from antisera against actin.

The cytoskeletal complex can be thought of as a three-tier system containing control systems, the cytoskeleton itself and finally cytoskeleton-associated proteins. The control systems may include proteins such as the centrosome-related antigen recognised by 5051 serum (Lloyd *et al.*, 1985) and calmodulin (Tiwari *et al.*, 1985). The cytoskeleton associated proteins interface between the cytoskeleton and other cellular components forming structural and perhaps functional links. Many such proteins are recognised in animal and microbial cells (Birchmeier, 1984) but few have been identified in plants.

How the cytoskeleton interacts with other cellular components within plant cells is, as yet, completely unknown at the molecular level. The cytoskeletal-associated proteins which are involved in such interactions may have diverged through evolution more than the main cytoskeletal proteins. No antisera raised, so far, against animal or lower plant cytoskeletal-associated proteins have been reported to cross-react with the corresponding higher plant proteins. Recently, antisera raised in rabbits against spinach calmodulin have been used to demonstrate the co-distribution of microtubules and calmodulin in the spindle and phragmoplasts of onion root tip cells but not with the interphase or PPB microtubules (Wick, Muto & Duniec, 1985). However, few such studies on cytoskeletal associated proteins have yet been undertaken. A fuller understanding of the cytoskeleton and its functions will have to await the availability of antisera raised specifically against plant proteins.

Acknowledgements

We are grateful to Dr C. W. Lloyd for critical reading of the manuscript; Peter Scott and Andrew Davies for photographic assistance and Anne Williams for preparation of the manuscript.

PROTOCOLS

I. Indirect immunofluorescent (IIF) staining of root tip meristem cells – rapid method

The aims of cell preparation for IIF include the maintenance of the integrity and shape of the cell, the fixation of the cytoskeleton in its native intercellular distribution and the permeabilization of both the cell wall and the cytoplasm to allow antibodies to diffuse freely in and out of cells. These aims are often intrinsically incompatible, so a compromise must be achieved. The first step, fixation, is followed in multicellular tissues, by the release of single cells, usually using cell wall degrading enzymes and EGTA. After squashing, or wall removal, plant cells are very fragile and must be supported during further handling.

1. Fix terminal 1–2 mm of primary root tips in 50 mM PIPES, 5 mM EGTA, 5 mM $MgSO_4$, pH 6.9 – 7.1 buffer containing 3.7% formaldehyde fixative[a].
2. Wash fixed roots in buffer minus fixative for 15 min (2 changes).
3. Treat with 2% cellulase (Onozuka R10) in above buffer for 20 min at room temperature[b].
4. Wash thoroughly in buffer minus enzyme for 30 min.
5. Squash out cells onto acid/acetone washed coverslips (poly-L-lysine coating optional), or onto the wells of Flow 8-well Multitest Slides[c].
6. Air dry at room temperature.
7. Incubate with first antibody – anti-yeast α tubulin monoclonal (Sera Labs). For ascites fluid at 1/200 dilution in PBS, 20 μl per well on Flow Slides, 40 min at 37 °C.
8. Wash thoroughly with PBS at room temperature (at least 3 changes).
9. Incubate with second antibody – Goat anti-rat FITC (Miles) at 1/150 dilution in PBS, 40 min at 37 °C.
10. Wash thoroughly with PBS adding 1 $\mu g\ cm^{-3}$ propidium iodide (Sigma) or 1 $\mu g\ cm^{-3}$ DAPI (Sigma) to slides/coverslips for 5 min, followed by several washes in PBS[d].
11. Mount slides/coverslips in glycerol: PBS 9:1, pH 8.0 – 9.0 containing either 1 mg cm^{-3} p-phenylene-diamine or 5% n-propyl gallate[e,f].

II. Immunofluorescent staining of intact moss plants

Plants are grown in cellophane-overlaid media for ease of removal from the agar medium (Doonan *et al.*, 1985).

1. Transfer the moss plant, using the underlying cellophane as support, to a dish containing 100 mM PIPES, 5 mM EGTA, 1 mM $MgSO_4$, 1% (v/v) DMSO, 0.01% (v/v) Triton X100 (pH 6.4), with 8% (w/v) formaldehyde as fixative.
2. Immerse for 20–30 min.
3. Wash fixed plants with 100 mM PIPES (pH 6.4), 5 mM EGTA, 1 mM $MgSO_4$ (MTSB) pH 6.4, for 5 min (2 changes).

4. Invert a freshly prepared glutaraldehyde-derivatized cover-slip (see Protocol IV) to cover the tissue and hold in place with a 20 g weight for 60 min.
5. Separate the coverslips with adhering tissue from the cellophane and place plant uppermost into 2% (w/v) Driselase (Sigma) solution containing 5 mM EGTA for 10–15 min[g].
6. Rinse twice in MTSB.
7. Place in MTSB containing 5% (v/v) DMSO and 3% (v/v) Nonidet NP40 for 1–4 h[h].
8. Wash free of detergent with MTSB (5/6 changes).
9. Stain as described for onion cells (Protocol I, steps 7–11).

III. Preparation of formaldehyde fixative

Formaldehyde is the preferred fixative for IIF of MTs although glutaraldehyde (Simmonds *et al.*, 1983) has been used. Various fixation protocols used for animal cells are reviewed by Osborn & Weber (1982). The formaldehyde fixative is made from solid para-formaldehyde as follows.

1. Heat 60 cm^3 distilled water to 60 °C and add 2–4 drops concentrated NaOH.
2. Add 4 g para-formaldehyde and stir until dissolved[i].
3. Allow to cool and, if necessary, filter through 0.45 μ Millipore filter.
4. Add 20 cm^3 buffer (five fold concentrated), adjust pH to between 6.4 and 6.9 with H_2SO_4/NaOH and make up to final volume of 100 cm^3.
5. Aliquot solution and store frozen (−20 °C)[j].

IV. Preparation of coverslips for cell attachment

Although sufficient adhesion may occur to clean glass, many cells will only stick to treated coverslips. Cells may be allowed to dry on acid cleaned glass-coverslips which also helps permeabilize their cytoplasm. Alternatively, they may be allowed to settle on poly-L-lysine coated coverslips or glutaraldehyde-derivatized coverslips. Cells which have not been air-dried may require further treatment in detergents or cold organic solvents to permeabilize their cytoplasm.

A. Polylysine-coated coverslips.
1. Clean coverslips in concentrated nitric acid.
2. Wash thoroughly in water.
3. Rinse in acetone and dry.
4. Layer about 400 μl of 0.1 mg cm^{-3} poly-L-lysine (Sigma) evenly over each coverslip and dry[j].
5. Rinse briefly in buffer before use.
6. Spread cell suspension over coverslip[k].

B. Glutaraldehyde-derivatized coverslips (modified from Aplin & Hughes, 1981).
1. Coverslips are cleaned in concentrated nitric acid.
2. Wash thoroughly in water.

3. Wash in absolute ethanol and dry.
4. Spread 200 μl of 4 aminopropyltriethoxysilane (Sigma) evenly over coverslip.
5. Incubate 2–3 min.
6. Rinse thoroughly in a large volume of water.
7. Cover in 0.2% (w/v) glutaraldehyde solution for 20–30 min.
8. Rinse free of glutaraldehyde with MTSB and use immediately.

Notes

a Onion roots require 50 min fixation.
b Time and concentration of enzyme is optimal for onion root-tips in Onozuka R10 Cellulase. This may be adjusted for different sources of enzymes, or other types of wall-lytic enzymes. Treatment should be such that the roots disperse into single cells with very little squashing; check by phase contrast.
c Dry down as thick a cell suspension as possible – most will be removed during washing (see Protocol IV for treatment of coverslips).
d DAPI may be added direct to the mountant at 0.1–0.5 μg/cm^{-3} as this stain only fluoresces when bound to DNA.
e Alternatively, use Mowiol 4–88 (Hoechst) containing the above anti-fade additives – gives a permanent preparation without sealing the coverslip with nail varnish. Citifluor anti-fade mountant (City University, London) also works.
f Stained cells in anti-fade mountant may be photographed using either FP4 or Pan F. Typical exposures for FITC are 30–90 s depending on the preparation and using Zeiss Planapo 40X objective with full aperture. Multiple exposures are possible allowing through focal series of single cells to be recorded.
g 2% Driselase for 10–15 min is optimal for moss protonemata. 5% (w/v) Driselase for 60 min is required for leaves and other plant tissues require other enzymes or combinations of enzymes to effect adequate cell wall removal. Treatments lasting longer than 60 min should have protease inhibitors added (e.g. 50 μg cm^{-3} leupeptin; 1 μg cm^{-3} phenyl methyl sulphonyl fluoride).
h The concentration and timing of detergent treatment must be optimized for tissues other than moss protonemata.
i Although 4% (w/v) formaldehyde is the commonly used concentration, fragile cells – tip growing root hairs (Lloyd & Wells, 1985), moss protonemata (Doonan *et al.*, 1985*a*) – are better preserved by 8% (w/v) formaldehyde solution.
j Formaldehyde fixative must either be used fresh or, if kept for later use, stored frozen at −20 °C.
k To increase adhesion coverslips may be allowed to dry.

KEY REFERENCES

De May, J., Lambert, A. M., Bajer, A. S., Moereman, M. & De Brabander, M. (1982). Visualization of microtubules in interphase and mitotic plant cells of *Haemanthus* endosperm with the immunogold staining method. *Proceedings of National Academy of Science, U.S.A.*, **79**, 1898–902.

Kilmartin, J. V., Wright, B. & Milstein, C. (1982). Rat monoclonal anti-tubulin antibodies derived by using a new non-secreting rat cell line. *Journal of Cell Biology*, **93**, 576–82.

Lloyd, C. W., Slabas, A. R., Powell, A. J., MacDonald, G. & Bradley A. R. (1979). Cytoplasmic microtubules in higher plants visualized with antitubulin antibodies. *Nature*, **279**, 239–41.

Porter, K. R. (1984). The Cytomatrix: a short history of its study. *Journal of Cell Biology*, **99**, 3s–12s.

Wick, S. M., Seagull, R. W., Osborn, M., Weber, K. & Gunning, B. E. S. (1981). Immunofluorescence microscopy of organised microtubule arrays in structurally stabilized meristematic plant cells. *Journal of Cell Biology*, **89**, 685–90.

REFERENCES

Aplin, J. D. & Hughes, R. C. (1981). Protein-derivatised glass coverslips for the study of cell to substratum adhesion. *Analytical Biochemistry*, **113**, 144–8.

Aubin, J. E. (1981). Immunofluorescence studies of cytoskeletal proteins during cell division. In *Mitosis/Cytokinesis*, ed. A. M. Zimmerman & A. Forer, pp. 211–244. New York: Academic Press.

Birchmeier, W. (1984). Cytoskeleton structure and function. *Trends in Biochemical Sciences*, 192–5.

Calarco-Gillam, P. D., Siebert, M. C., Hubble, R., Mitchison, T. & Kirschner, M. (1983). Centrosome development in early mouse embryos as defined by an auto-antibody against pericentriolar material. *Cell*, **35**, 621–9.

Coss, R. A. & Pickett-Heaps, J. D. (1974). The effects of isopropyl N-phenyl carbamate on the green alga *Oedogonium cardiacum*. I. Cell division. *Journal of Cell Biology*, **63**, 84–97.

Clayton, L. & Lloyd, C. W. (1984). The relationship between the division plane and spindle geometry in *Allium* cells treated with CIPC and griseofulvin; an anti-tubulin study. *European Journal of Cell Biology*, **34**, 248–53.

Clayton, L. & Lloyd, C. W. (1985). Actin organization during the cell cycle in meristematic plant cells: actin is present in the cytokinetic phragmoplast. *Experimental Cell Research*, **156**, 231–8.

Clayton, L., Black, C. M. & Lloyd, C. W. (1985). Microtubules in higher plant cells identified by an autoantibody against pericentriolar material. *Journal of Cell Biology*, **101**, 319–23.

Dawson, P. J., Hulme, J. S. & Lloyd, C. W. (1985). Monoclonal antibody to intermediate filament antigen cross-reacts with higher plant cells. *Journal of Cell Biology*, **100**, 1793–8.

DeMaggio, A. E. & Stetler, D. A. (1977). Protonemal organization and growth in the moss *Dawsonia superba*: ultrastructural characteristics. *American Journal of Botany*, **64**, 449–54.

Dickinson, H. G. & Sheldon, J. M. (1984). A radial system of microtubules extending between the nuclear envelope and the plasma membrane during early halophase in flowering plants. *Planta*, **161**, 86–90.

Doonan, J. H., Cove, D. J. & Lloyd, C. W. (1985). Immunofluorescence microscopy of microtubules in intact cell lineages of the moss *Physcomitrella patens*. I. Normal and CIPC-treated tip cells. *Journal of Cell Science*, **75**, 131–47.

Doonan, J. H., Lloyd, C. W. and Duckett, J. G. (1986). Anti-tubulin antibodies locate the blepharoplast during spermatogenesis in the fern *Platyzoma microphyllum* R.BR.: a correlated immunofluorescence and electron-microscopic study. *Journal of Cell Science* (in press).

Dustin, P. (1984). *Microtubules*, 2nd edition. Springer-Verlag: Berlin, Heidelberg, New York, Tokyo.

Falconer, M. M. & Seagull, R. W. (1985). Immunofluorescent and Calcofluor White staining of developing tracheary elements in *Zinnia elegans* L. suspension cultures. *Protoplasma*, in press.

Franke, W. W., Seib, E., Osborn, M., Weber, K., Herth, W. & Falk, H. (1977). Tubulin-containing structures in the anastral mitotic apparatus of endosperm of the plant *Leucojum aestevum* as revealed by immunofluorescence microscopy. *Cytobiology*, **15**, 24–48.

Gunning, B. E. S. (1982). The cytokinetic apparatus: its development and spatial regulation. In *The Cytoskeleton in Plant Growth and Development*, ed. C. W. Lloyd, pp. 229–94. London: Academic Press.

Gunning, B. E. S., Hardham, A. R. & Hughes, J. E. (1978). Pre-prophase bands of microtubules in all categories of formative and proliferative cell division in *Azolla* roots. *Planta*, **143**, 145–60.

Hawes, C. R. & Horne, J. C. (1985). Polyethylene glycol embedding of plant tissues for transmission electron microscopy. *Journal of Microscopy*, **137**, 35–45.

Jensen, L. C. W. (1981). Division, growth and branch formation in the protonemata of the moss *Physcomitrium turbinatum*; studies of sequential cytological changes in living cells. *Protoplasma*, **107**, 301–17.

Knox, R. B. (1982). Methods of locating and identifying antigens in plant tissues. In *Techniques in Immunocytochemistry*, vol. **1**, ed. G. R. Bullock & P. Petrusz. London: Academic Press.

Ledbetter, M. C. & Porter, K. R. (1963). A "microtubule" in plant cell fine structure. *Journal of Cell Biology*, **19, 239–50**.

Lloyd, C. W. (1983). Helical microtubular arrays in onion root hairs. *Nature*, **305**, 311–13.

Lloyd, C. W. (1984). Toward a dynamic helical model for the influence of microtubules on wall patterns in plants. *International Review of Cytology*, **85**, 1–51.

Lloyd, C. W., Clayton, L., Dawson, P. J., Doonan, J. H., Hulme, J. S., Roberts, I. & Wells, B. (1985). The cytoskeleton underlying side walls and cross walls in plants: molecules and macromolecular assemblies. *Journal of Cell Science* (in press).

Lloyd, C. W., Slabas, A. R., Powell, A. J. & Lowe, S. B. (1980). Microtubules protoplasts and plant cell shape. An immunofluorescent study. *Planta*, **147**, 500–6.

Lloyd, C. W. & Wells, B. (1985). Microtubules are at the tips of root hairs and form helical patterns corresponding to inner wall fibrils. *Journal of Cell Science*, **75** (in press).

Marchant, H. J. (1978). Microtubules associated with the plasma membrane isolated from protoplasts of the green alga *Mougeotia*. *Experimental Cell Research*, **115**, 25–30.

Newcomb, E. H. & Bonnett, H. T. (1965). Cytoplasmic microtubule and wall microfibril orientation in root hairs of radish. *Journal of Cell Biology*, **27**, 575–89.

Oliver, J. M., Krawiec, J. A. & Berlin, R. D. (1978). A carbamate herbicide causes microtubule and microfilament disruption and nuclear fragmentation in fibroblasts. *Experimental Cell Research*, **116**, 229–37.

Olmstead, J. B., Cox, J. V., Asnes, C. F., Parysek, L. M. & Lyon, H. D. (1984). Cellular regulation of microtubule organization. *Journal of Cell Biology*, **99**, 28s–32s.

Osborn, M. & Weber, K. (1982). Immunofluorescence and immunocytochemical procedures with affinity purified antibodies: tubulin containing structures. In *Methods of Cell Biology*, vol. **24**, The Cytoskeleton Part A, ed. L. Wilson. London: Academic Press.

Pickett-Heaps, J. D. (1967). The effects of colchicine on the ultrastructure of dividing plant cells xylem wall differentiation and distribution of cytoplasmic microtubules. *Developmental Biology*, **15**, 206–36.

Pickett-Heaps, J. D. (1969). Preprophase microtubules and stomatal differentiation: some effects of centrifugation on symmetrical and asymmetrical cell division. *Journal of Ultrastructural Research*, **27**, 24–44.

Pickett-Heaps, J. D. & Northcote, D. H. (1966). Organization of microtubules and endoplasmic reticulum during mitosis and cytokinesis in wheat meristems. *Journal of Cell Science*, **1**, 109–20.

Powell, A. J., Lloyd, C. W., Slabas, A. R. & Cove, D. J. (1980). Demonstration of the microtubular cytoskeleton of the moss *Physcomitrella patens* using antibodies against brain tubulin. *Plant Science Letters*, **18**, 401–4.

Pruss, R. M., Mirsky, R., Raff, M. C., Thorpe, R., Dowding, A. J. & Anderton, B. H. (1981). All classes of intermediate antibody share a common antigenic determinant defined by a monoclonal antibody. *Cell*, **27**, 419–28.

Roberts, I. N., Lloyd, C. W. & Roberts, K. (1985). Ethylene-induced microtubule re-orientations: mediation by helical arrays. *Planta*, **164**, 439–47.

Robinson, D. G. & Quader, H. (1982). The microtubule-microfibril syndrome. In *The Cytoskeleton in Plant Growth and Development*, ed. C.W. Lloyd, pp. 109–26. London: Academic Press.

Schmiedel, G. & Schnepf, E. (1980). Polarity and growth of caulonema tip cells of the moss *Funaria hygrometrica*. *Planta*, **147**, 405–13.

Schnepf, E. (1973). Mikrotubulus-Anordnung und-Umordnung Wandbildung und Zellmorphogenese in jungen *Sphagnum* Blättchen. *Protoplasma*, **78**, 145–73.

Seagull, R. & Heath, I. B. (1980). Identification of microtubular structure microtubule arrays in radish root hair. *Protoplasma*, **103**, 205–29.

Schmit, A. C., Vantard, M., de Mey, J. & Lambert, A. M. (1983). Aster-like microtubule centres establish spindle polarity during interphase-mitosis transition in higher plant cells. *Plant Cell Reports*, **3**, 285–88.

Simmonds, D. H., Seagull, R. W. & Setterfield, G. (1985). Evaluation of techniques for immunofluorescent staining of MTs in cultured plant cells. *Protoplasma*, in press.

Simmonds, D., Setterfield, G. & Brown, D. L. (1983). Organization of microtubules in dividing and elongating cells of *Vicia hajastana* Grossh. in suspension culture. *European Journal of Cell Biology*, **32**, 59–66.

Steen, D. A. & Chadwick, A. V. (1981). Ethylene effects in pea stem tissue, evidence of microtubule mediation. *Plant Physiology*, **67**, 460–6.

Steinart, P. M., Jones, R. C. R. & Goldman, R. D. (1984). Intermediate filaments. *Journal of Cell Biology*, **99**, 22s–8s.

Stossel, T. P. (1984). Contribution of actin to the structure of the cytoplasmic matrix. *Journal of Cell Biology*, **99**, 15s–21s.

Tiwari, S. C., Wick, S. M., Williamson, R. E. & Gunning, B. E. S. (1985). Cytoskeleton and integration of cellular function in cells of higher plants. *Journal of Cell Biology*, **99**, 63s–9s.

Van der Valk, P., Rennie, P. J., Connolly, J. A. & Fowke, J. C. (1980). Distribution of cortical microtubules in tobacco protoplasts. An immunofluorescence microscopic and ultrastructural study. *Protoplasma*, **105**, 27–43.

Wada, M., Mineyuka, Y., Kadota, A. & Furuya, M. (1980). The changes in nuclear position and distribution of circumferentially aligned cortical microtubules during the progression of the cell cycle in *Adiantum* protonemata. *Botanical Magazine, Tokyo*, **93**, 237–45.

Weber, K., Osburn, M., Franke, W. W., Seib, E., Schneer, V. & Herth, W. (1977). The organization of cortical microtubules, in diverse plant and animal cells by immunological cross-reaction revealed by immunofluorescence microscopy using antibody against tubulin from porcine brain. *Cytobiologie*, **15**, 285–302.

Wick, S. M. & Duniec, J. (1983). Immunofluorescence microscopy of tubulin and microtubule arrays in plant cells. I. Pre-prophase band development and concomitant appearance of nuclear envelope-associated tubulin. *Journal of Cell Biology*, **97**, 235–43.

Wick, S. M. & Duniec, J. (1984). Immunofluorescence microscopy of tubulin and microtubule arrays in plant cells. II. Transition between pre-prophase band and mitotic spindle. *Protoplasma*, **122**, 45–55.

Wick, S. M., Muto, S. & Duniec, J. (1985). Double immunofluorescence labelling of calmodulin and tubulin in dividing plant cells. *Protoplasma*, **126**, 198–206.

Wick, S. M., Seagull, R. W. Osborn, M., Weber, K. & Gunning, B. E. S. (1981). Immunofluorescence microscopy of organized microtubule arrays in structurally stabilised meristematic plant cells. *Journal of Cell Biology*, **89**, 685–90.

A.C. CUMING, R.S. WILLIAMS
and J.V. CULLIMORE

The use of antibodies in molecular biology

Introduction

"Molecular Biology" is a diffuse descriptive term which covers a variety of disciplines. However, the aims of molecular biologists may be succinctly stated as the attempt to define biological phenomena by a description of the underlying molecular events – to understand the reification of genetic information in structures which participate in the maintenance of cellular functions. These aims are being achieved through studies of the molecular structure of cellular components and through studies of the process of information transfer itself: the process of gene expression.

The mammalian immune system is a highly developed self-defence mechanism. The introduction of foreign matter (antigens) to animals elicits a vigorous response directed towards the inactivation of these antigens. This response includes the production of a class of proteins, the immunoglobulins (antibodies), which recognize individual determinants on the surface of the antigen and react with them to bring about their immobilization. Subsequently, they are destroyed by other participants in the immune response. The exquisite specificity with which antibodies recognize particular determinants makes them ideal tools for the molecular biologist. Their specificity permits them to be used as probes for molecular structure and function and also for the recognition and isolation of individual molecular species from the cellular milieu.

It is this latter use to which we address our discussion in this article. We shall describe, with reference to work in progress in our laboratories, how immunochemical methods can be used in conjunction with recombinant DNA techniques to determine the means by which the expression of specific genes is modulated during plant development.

Fig. 1 illustrates, in simplified form, the stages which occur during the expression of eukaryotic genes. It is clear that the cascade of information transfer is complex, with opportunities for regulatory events to occur at each level in the pathway. Prior to transcription, the state of the DNA may dictate its activity. The expression of a gene may first require changes in the chromatin conformation, resulting from altered associations between

DNA and DNA-binding proteins, which permit access to polymerases. Subsequently, DNA sequence amplification, sequence rearrangement or base modification (e.g. methylation/demethylation) may be necessary for expression. In addition to regulation of this nature, it has become increasingly apparent that the transcription of individual genes at specific stages of development, or in response to environmental stimuli can be a consequence of the association of the structural sequence with a regulatory promoter sequence "upstream" of the transcriptional initiation site, or with an "enhancer" sequence, which may be located either 3′ or 5′ to the coding sequence.

Following transcription, processing of the primary transcript occurs to produce the translatable messenger RNA (mRNA). Processing events can include the removal of non-coding intervening sequences ("splicing"), base modification (e.g. by methylation) and polyadenylation, both of which enhance the stability of the messenger, and 5′-terminal "capping" which confers resistance to attack by exonucleases and increases the efficiency of the initiation of translation. During the production of a messenger, the transcript also becomes associated with proteins, to form the mature

Fig. 1. Expression of an "eclectic" eukaryotic gene. A schematic representation of the expression of a plant gene. The filled blocks represent sequences which encode the mature polypeptide product. Abbreviations are as follows:
DR – promoter responsible for developmental specificity of expression.
P – promoter ("TATA" box).
TS – transcription start site.
ls – 5′ untranslated leader sequence in mRNA.
TP – transit peptide (for transport of organellar proteins).
I1,2 & 3 – non-coding intervening sequences ("introns").
3′ unt – 3′ untranslated sequence in mRNA: may differ in differentially expressed members of the same gene family.
A – sequence specifying polyadenylate addition.
TT – transcription termination sequence.
En – enhancer sequence.

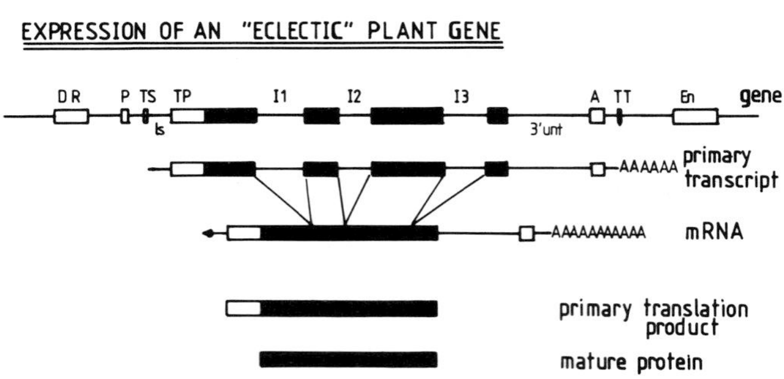

messenger ribonucleoprotein particle (mRNP). These proteins contribute to the stability of the mRNA and may also regulate its recruitment into polyribosomes. Thus the translational activity of a messenger is not simply a function of its abundance, but is also a consequence of its degree of base modification, secondary structure, protein associations and resistance to degradation within the cell.

Following translation, the initial polypeptide product may not be identical with its active form. For the gene product to become functionally active, it may have to be transported to a particular subcellular destination, e.g. within an organelle, or to an extracellular location. Frequently, such polypeptides contain the information for this targeting in the form of amino-terminal "signal" or "transit" peptides. These amino-terminal extensions are removed by the action of endonucleases during the act of transport. The activity of individual polypeptide gene products may further be modulated by secondary modifications, such as phosphorylation, glycosylation, ribosylation, etc., or by posttranslational proteolytic cleavage, such as occurs in the synthesis and packaging of seed storage proteins. Finally, the gene product may interact with the products of other genes, as in the assembly of multimeric enzymes. Between DNA sequence and functional gene product lie a myriad of possible intermediate forms. To trace the path by which the expression of an individual gene is achieved, it is therefore necessary to identify each of these intermediates and to do so among the "background" expression of up to perhaps 10^5 other genes. For this task, specific probes are required for the recognition of the polynucleotide or polypeptide of interest. Cloned DNA sequences provide probes for DNA and RNA sequences and mono-specific antibodies are used to identify individual polypeptides.

The use of antibodies to identify polypeptides

The analysis of complex mixtures of proteins is most conveniently achieved by polyacrylamide gel electrophoresis (PAGE). When a unidimensional separation is performed (e.g. by sodium dodecylsulphate (SDS)–PAGE)), the number of polypeptides resolved is of the order of 10^2 (Laemmli, 1970). Two-dimensional separations achieve far higher resolution, up to 10^4 polypeptides being resolved (O'Farrell, 1975). The identification of individual polypeptides within these complex electrophoretic patterns may be achieved by the use of an antibody to precipitate, with high specificity, a radiolabelled antigen. For such a procedure to succeed, it is necessary for the mixture of polypeptides being assayed to be radiolabelled to high specific activity. This is frequently difficult to achieve by the application of radioactive isotopes to plant tissue *in vivo* although, as we shall demonstrate, the technique of immunoprecipitation is well suited to the analysis of polypeptide mixtures labelled *in vitro*. An alternative strategy has recently been developed to identify unlabelled polypeptides in electrophoretic separations by immunoassay and is based on the ability of proteins to bind to inert matrices such as nitrocellulose or nylon membranes. The procedure involves the electrophoretic transfer of polypeptides, resolved by PAGE, from a polyacrylamide gel to the membrane, where they become immobilized. The membrane thus forms a "filter replica" of the gel, with the bound polypeptides in a conformation accessible to interaction with specific antibodies (Towbin, Staehelin & Gordon, 1979; Vaessen, Kreike & Groot, 1981). Incubation of the filter with a specific antiserum will result in the antibody in that serum binding to the antigen in its position on the filter. When unbound antibody is removed by washing, the position of the antibody-antigen complex can be located on the filter by locating the position of the bound antibody. This can be achieved either by binding a peroxidase-linked second antibody (e.g. goat anti-rabbit immunoglobulin (IgG)) (Towbin *et al.*, 1979) or by binding ^{125}I-Protein A (Vaessen *et al.*, 1981). (Protein A isolated from *Staphylococcus aureus* cells specifically binds IgG molecules, the main component of antiserum.) This "immunoblotting" technique is analogous to the blot transfer methods used for the analysis of nucleic acids (Southern, 1975), and is therefore frequently (if frivolously) referred to as "Western" blotting. (See Roberts, this volume, for the use of filters in dot immunobinding for monoclonal antibody (McAb) screening.)

In Fig. 2, we show how such an assay has been used to screen a series of McAb raised against polypeptides contained in photosynthetic membranes. The photosystem I (PS I) particle of pea chloroplasts is a chlorophyll-lipoprotein complex containing 16 stainable polypeptides (Mullet, Burke & Arntzen, 1980; see also Fig. 2, track 1). Entire PS I particles were used to immunize mice, whose spleen cells were fused with myeloma cells to generate a collection of hybridoma cultures (Koehler & Milstein, 1975) which secreted antibodies to the individual polypeptides of the complex. Fig. 2, track 2 illustrates the binding of antiserum to each polypeptide of the PS I particle. In this case, the serum used for the immune reaction was obtained by bleeding the mouse whose spleen was used to generate the hybridomas. Tracks 3 and 4 demonstrate the binding of the antibodies secreted by two of the hybridoma lines. In track 3, the antibody specifically recognized polypeptide P-2 of the PS I-associated light-harvesting complex (LHC-1) and in track 4, the antibody recognized polypeptide P-3 of the LHC-1 and its dimeric form. The protocol by which the hybridoma lines were screened by the immunoblot method is illustrated in cartoon form in

Fig. 2. Immunoblot identification of polypeptides of the pea chloroplast photosystem I complex.

Track 1 – electrophoretic separation of PS I-polypeptides; Coomassie Blue stain.

Track 2 – filter replica of PS I polypeptides reacted with polyclonal anti-PS I serum.

Track 3 – filter replica of PS I polypeptides reacted with monoclonal antibody to polypeptide P2 of LHC-1 (the PS I-associated light-harvesting complex).

Track 4 – filter replica of PS I polypeptides reacted with monoclonal antibodies to polypeptide P3 of LHC-1. In this gel system, the slower migrating band corresponds to a dimer of P3, which is incompletely dissociated. The polypeptide at 110kd corresponds to the undenatured PS I reaction centre polypeptide, and that at 68kd to the fully denatured form. LHC-2 indicates the position of polypeptides of the light-harvesting chlorophyll *a*/*b* – protein complex, which is associated with both PS I and PS II.

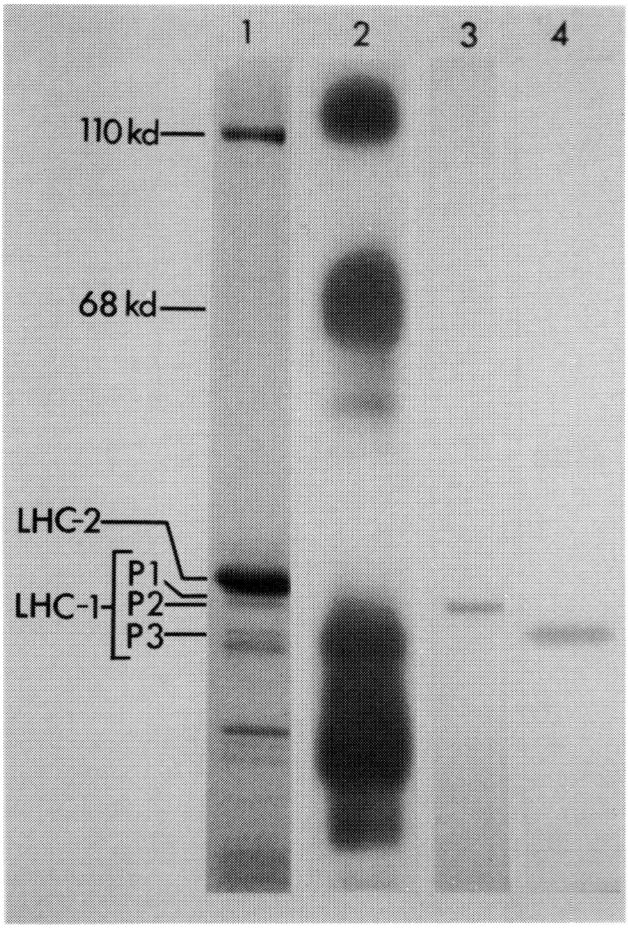

Fig. 3 and the technical details are presented in the protocols at the end of the chapter.

It is important to note that the use of the immunoblot technique coupled with the production of McAb obviates the need to isolate a single antigenic polypeptide in pure form for the immunization of animals. This makes the procedure particularly useful for the study of polypeptides which comprise only a small proportion of the cellular protein content, or which are otherwise difficult to purify. (See Roberts, this volume and Brewin *et al.*, this volume for examples.)

Fig. 3. Immunoblot screening of monoclonal antibodies.

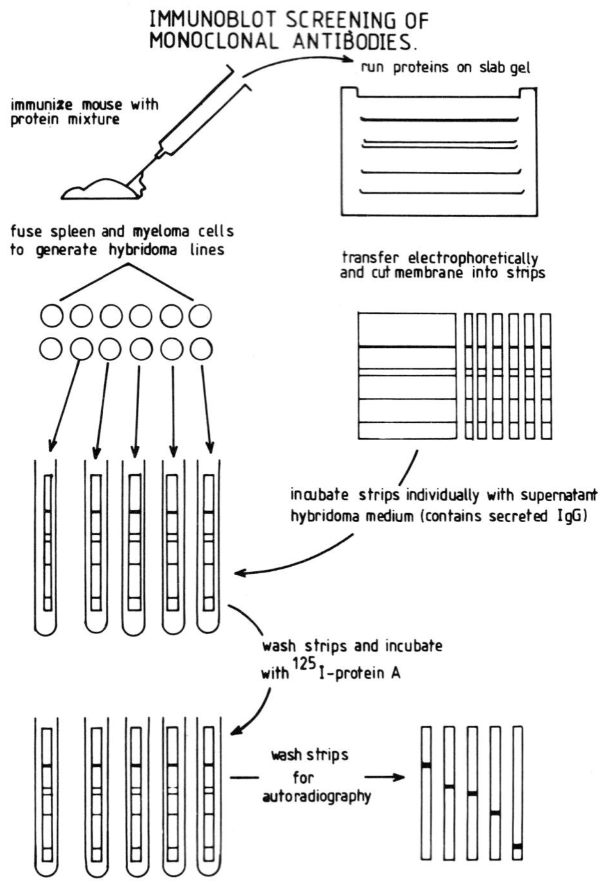

The use of antibodies to assay messenger RNA

Although immunoblotting is a powerful technique for the identification of small quantities of individual polypeptides in complex mixtures, it is not suitable for the identification of transient forms of these polypeptides such as precursors. Proteins which pass through membranes or those integrated into the structure of membranes are usually synthesized in the form of a precursor. These precursors are difficult to identify, *in vivo*, because they are usually rapidly transported and processed to the mature form. However, when they are synthesized *in vitro* in the absence of their "target membrane", the precursors are not processed and may be detected by immunoprecipitation. We have used this technique to identify the precursors of the major polypeptides of the light-harvesting chlorophyll *a*/*b* – protein complex (LHC-2) of pea chloroplast membranes (Cuming & Bennett, 1981). These polypeptides (M*r* = 26,000 and 24,000) were isolated from SDS-polyacrylamide gels and used to immunize rabbits. When mRNA isolated from pea leaves was translated in a wheat-germ cell-free system (Roberts & Paterson, 1973) and the translation products reacted with the anti-LHC-2 antibodies, two polypeptides of M*r* = 32,000 and 30,000 were specifically precipitated (Fig. 4, tracks 1 and 2, see Protocols for details of the method). The identity of these precursors was confirmed by peptide mapping and the specificity of the immunoprecipitation also permitted the procedure to be used quantitatively to estimate the levels of LHC-2 mRNA present in pea leaves under different conditions of illumination by measuring the recovery of the radiolabelled translation product in the immunoprecipitate (Cuming & Bennett, 1981). Subsequently, these estimates of the messenger level were confirmed by direct measurement, using a cloned DNA probe (Jenkins, Hartley & Bennett, 1983). The identification of this DNA probe was considerably facilitated by the availability of the LHC-2 – specific antibody.

In order to use a cloned DNA sequence to analyse the expression of a particular gene, it is first necessary to demonstrate that the cloned sequence encodes the appropriate gene product. If the amino acid sequence of the polypeptide is known, then the identity of a cloned DNA sequence may be confirmed by DNA sequence analysis. However, this situation seldom obtains and it is more usual for the cloned sequence to be used to select messenger RNA, by hybridization, which subsequently can be analysed by translation, *in vitro* (Bedbrook, Smith & Ellis, 1980). The product of this "hybrid-selected translation" can then be identified by immunoprecipitation.

Identification of a cloned sequence encoding the largest polypeptide of LHC-2 is illustrated in Fig. 4 (tracks 3 and 4). This identification was facilitated by two additional factors. Firstly, the LHC-2 mRNA was subject to developmental regulation, being tenfold more abundant in green leaves than in etiolated tissue. Secondly, the messenger was relatively abundant in green leaves, comprising approx. 1–2% of the total messenger fraction. The first factor permitted a preliminary screening by colony hybridization (Grunstein & Hogness, 1975); only those clones which hybridized strongly with mRNA from green leaves and weakly with mRNA from etiolated tissue being chosen for further screening. The second factor permitted efficient hybridization between the cloned DNA and the LHC-2 mRNA. These factors are both important because the technique of hybrid-selected translation is both technically demanding and laborious. It is necessary to isolate individual plasmids in pure form, linearize and denature them, and bind them to a suitable support medium (usually nitrocellulose, or diazotized

Fig. 4. Immunoprecipitation of LHC-2 precursor polypeptides.
Track 1 – total translation products of green pea leaf poly(A)-containing RNA.
Track 2 – translation products of green pea leaf poly(A)-containing RNA immunoprecipitated by antibodies to the 26kd and 24kd LHC-2 polypeptides.
Track 3 – translation products of mRNA selected by hybridization to a cDNA clone (pFab 31) encoding the 26kd LHC-2 polypeptide.
Track 4 – translation products of mRNA from *Xenopus laevis* ovaries mixed with translation products of pFab 31-selected mRNA and immunoprecipitated with anti-LHC-2 antibodies.

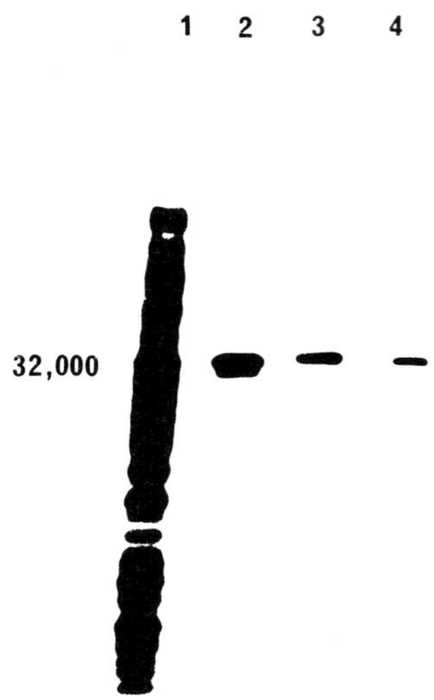

paper). This achieved, the individual filter-bound plasmids are incubated with a quantity of mRNA to permit hybridization of complementary sequences, washed thoroughly to remove non-specifically bound RNA and treated to elute the hybridized mRNA. This is recovered by ethanol-precipitation and translated *in vitro* for subsequent assay by immuno-precipitation.

The use of antibodies to purify messenger RNA

In many cases these favourable conditions will not obtain. Frequently it is a desired object to isolate a cloned DNA sequence encoding a polypeptide whose mRNA is neither subject to marked developmental regulation, nor is of great abundance within the cellular messenger population. In such a case an alternative means of obtaining a specific probe must be essayed. One such approach we have utilized successfully is the isolation of a pure messenger species by the immunopurification of polysomes.

Immunopurification of polysomes is a technique designed to purify mRNAs for proteins to which antisera have been raised. Although demonstrated as early as 1961 (Cowie *et al.*, 1961), only in the past few years has this method been developed to be applicable for purifying messengers of different relative abundances isolated from a wide variety of species and tissues. The significance of this technique is that the purified mRNA may then be used as a template for cloning, or to provide a radioactive probe to identify the clone of interest in a "library" of copy (cDNA) or genomic DNA clones. In this way the antiserum is used indirectly to identify the genes coding for the antigenic protein.

The technique relies on the antibodies reacting with their antigens on the nascent polypeptides of polysomes (Fig. 5). The antibody-bound polysomes can then be separated free of the unreacting polysomes and the specific mRNA isolated from them. Originally, the separation of the antibody-bound polysomes relied on either immunoprecipitation or on binding of the antibody-polysomes complex to an antigen column (Palmiter, Palacios & Schimke, 1972; Schimke *et al.*, 1974). However, this proved to be successful only for purifying mRNAs encoding abundant proteins, such as ovalbumin (Palacios *et al.*, 1973) and immunoglobulin (Schechter, 1973). A recent improvement of this technique uses Protein-A Sepharose to isolate the antibody–polysome complex by affinity chromatography (Shapiro & Young, 1981; see also Fig. 5). Unbound polysomes are washed from the column and the specifically bound fraction is eluted following dissociation with EDTA. The messenger RNA from the dissociated polysomes, being polyadenylated at the 3′-terminus, is recovered by chromatography over oligo-d(T) cellulose. The messenger eluted from this column is recovered by ethanol precipitation, and its purity is established by translation, *in vitro*.

By using these modifications specific messengers encoding seed lectin and Kunitz trypsin inhibitor (Vodkin, 1981) and the enzyme glutamine synthetase (Cullimore & Miflin, 1983) have been isolated from plant tissue. In the last example, glutamine synthetase (GS) mRNA was purified to near homogeneity from polysomes of *Phaseolus vulgaris* root nodules. When translated, this mRNA directed the synthesis of a single polypeptide which comigrated with purified GS when analysed by SDS-PAGE (Fig. 6). This purified GS mRNA was then radioactively labelled and used to screen a "library" of 1,100 cDNA clones prepared from *P. vulgaris* root nodule mRNA. A single clone in this "library" was identified as containing sequences encoding GS (Cullimore *et al.*, 1985). The methods used in this study are described in the Protocols. This technique may also be used with McAb and it has recently enabled the purification of messenger sequences of relatively low (0.1%) abundance (Korman *et al.*, 1982; Kraus & Rosenberg, 1982).

Despite these recent successes, several caveats should be observed. Firstly, the production of antiserum is often a limiting step, particularly for

Fig. 5. Immunopurification of polysomes.

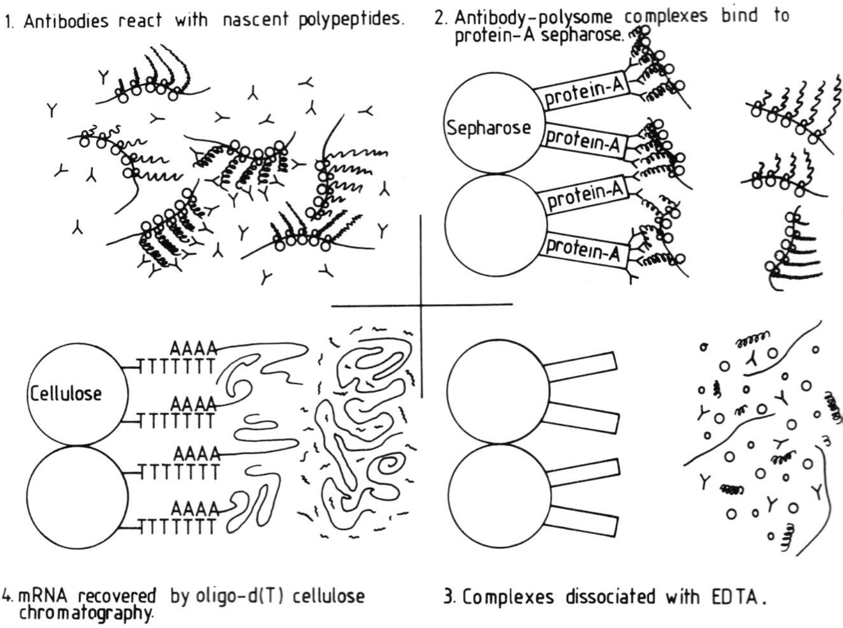

proteins of low abundance or for those which are otherwise difficult to purify. This problem may be overcome by the use of McAb which can be produced in bulk in cell cultures. Secondly, even when antiserum has been obtained, it may not cross-react with the nascent polypeptide. Thirdly, the isolation of polysomes translating the mRNA of interest may also pose problems. Time is well spent in screening for plant material where the desired mRNA is of the highest possible relative abundance. With these limitations in mind, we present the protocol by which the GS mRNA was immunopurified at the end of this chapter.

Direct screening of cDNA clones with antibodies

Antibodies may also be used for the direct screening of a cDNA "library" when the vector-host combination chosen is capable of expressing

Fig. 6. Analysis of mRNA prepared by immunopurification of polysomes using anti-glutamine synthetase (GS) IgG.
Track 1 and 2 – products of endogenous protein synthesis in reticulocyte lysates.
Track 3 – translation products of total polysomal RNA from *Phaseolus vulgaris* root nodules.
Track 4 – translation products of total polyadenylated RNA from root nodules.
Track 5 – translation products of polyadenylated RNA isolated from polysomes immunopurified with anti-GS IgG.
Track 6 – purified GS.

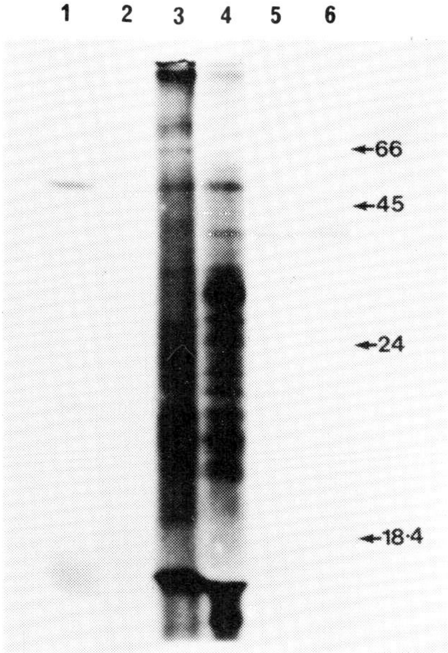

the cloned DNA sequence as a polypeptide. Such techniques have been successfully carried out in "model" systems but have not yet achieved widespread acceptance. These methods depend on *in situ* lysis of recombinant transformants to release the polypeptide which may then be immobilized for detection with a labelled antibody. Various methods for the immobilization of the antigen have been described, including binding to nitrocellulose filters (Young & Davis, 1983), covalent attachment to cyanogen bromide activated paper (Kemp & Cowman, 1981) and reaction with specific antibodies bound to polyvinyl discs (Broome & Gilbert, 1978).

In order for such methods to succeed, several prerequisites must first be satisfied. The DNA sequence must be introduced to a vector in such a way as to ensure its expression. For cloning in bacterial systems, it is therefore necessary to insert the sequence "downstream" of a suitable bacterial promoter. Because "foreign" proteins made in bacterial cells are frequently rapidly degraded, protocols based on the immunodetection of recombinants have therefore usually included the construction of fusion proteins – the DNA to be introduced to the vector is cloned at a site in the structural region of a bacterial gene, so that the N-terminal portion of the polypeptide product is bacterial and the C-terminal portion is "foreign". For an expressed polypeptide to be successfully recognized by an antibody, it is necessary that the introduced DNA sequence is cloned in the correct orientation and in the correct reading frame relative to the bacterial gene. This means that only one-sixth of the recombinants will be likely to produce a recognizable protein. It is therefore advisable, if adopting this cloning strategy, to utilize a system which will yield a large number of recombinants. It is probable that the most successful vectors for this purpose will be akin to the phage λ gt 11, described by Young and Davis (1983). It is also important to recognize that the detectability of the fusion product will further depend on the nature of the cDNA sequence that has been introduced. Since the antibody-antigen association is based on the recognition by the antibody of the conformation of the antigen, fusions in which the antigenic polypeptide constitutes only a small proportion of the hybrid protein are less likely to be detected. Although direct screening of recombinants is still relatively restricted in application, it is to be hoped that further "tuning" of the cloning vehicles will result in such protocols becoming more generally used.

PROTOCOLS

I. Immunoblotting for the screening of monoclonal antibodies

1. Resolve the protein mixture by SDS-PAGE (Laemmli, 1970) using a slab gel[a].
2. Cut a piece of nitrocellulose membrane (45 μm pore size) to match the shape of the gel and pre-soak it in transfer buffer – 192 mM glycine; 25 mM Tris-HCl; 20% methanol; pH 8.3).
3. Lay the membrane over the gel, being careful to avoid air bubbles and sandwich the gel and membrane between two layers of a soft porous material, e.g. nylon scouring pads or foam plastic.
4. Immerse the whole assembly in transfer buffer, so as to pass an electric current transversely through the gel. Several commercially made apparatuses are obtainable for this purpose[b].
5. After transfer, cut the filter as required and preincubate for 30 min in phosphate buffered saline (PBS – 10 mM Na phosphate; pH 7.2; 150 mM NaCl) containing 8% (w/v) bovine serum albumin (BSA) with shaking at room temperature for 30 min[c].
6. Incubate filters with antibody in PBS and under the same conditions as step 5 for at least 2 hours (it is often convenient to carry out this step overnight). Keep the volume of the incubation as low as possible, whilst maintaining the filter immersed[d].
7. Wash the filter 3–4 times with PBS (no BSA) to remove unbound antibody.
8. Incubate the filter in PBS plus BSA containing 10^6 cpm ^{125}I-labelled protein A for 2 hours, keeping the volume as low as is commensurate with the immersion of the filter by freely-moving liquid[e].
9. Remove unbound protein A by serial washing in PBS containing 1% Triton X-100.
10. Dry the filter and expose it to X-ray film to locate the bound radioactivity.

II. Purification of antibodies for immunoprecipitation

Immunoprecipitation of translation products can be carried out with relatively impure serum preparations. However, optimization of the conditions is better achieved if purified IgG is used. For the immunopurification of polysomes, it is essential to use a pure ribonuclease-free IgG fraction. We have used affinity purification (steps 1–4; Hjelm, Hjelm & Sjoquist, 1972) or ion exchange chromatography (steps 5–7; Palacios, *et al.*, 1972).

EITHER

1. Crude antiserum (5 cm^3) is passed over a column of Protein A Sepharose (Pharmacia, 1.7 cm^3).
2. The column is washed with PBS.
3. Elute and precipitate the IgG by the addition of ammonium sulphate to 40%.

4. Recover the IgG by centrifugation, dissolve the precipitate in PBS and dialyse against PBS at 0 °C.
   OR
5. Prepare a column 12 cm × 1 cm of DEAE cellulose (DE-52) overlaid with an equal quantity of CM cellulose (CM-52).
6. Equilibrate and sterilize in PBS buffer.
7. Apply the IgG in PBS and elute in the void volume. This fraction is electrophoretically pure, ribonuclease-free IgG. It can be stored frozen at −20 °C, and thawed without loss of activity.

III. Immunoprecipitation of polypeptides from cell free translations

Translation of mRNA is carried out in small volumes of wheat-germ S-30 extract (Roberts & Paterson, (1973) or rabbit reticulocyte lysate (Pelham & Jackson, 1976). The following protocols apply to a 20 μl reaction mixture.

1. Remove 2 × 1 μl aliquots to determine incorporation of radioactivity into TCA-insoluble material.
2. Add 1 μl RNase A (50 mg cm^3), 1 μl RNase T1 (50 units) and digest at 37 °C for 30 min[f].
3. Add 2 μl 10% SDS and boil for 2 min.
4. Dilute the mixture to 200 μl by the addition of PBS containing 1% BSA[g].
5. Add 10 μg IgG and incubate at 37 °C for 2 hours.
6. Weigh out 10 mg Protein A Sepharose (Pharmacia) and swell with 0.8 cm^3 PBS containing 1% BSA and 1% NP-40.
7. Add the mixture from Step 6 to the immunoprecipitation reaction mixture in a 1.5 cm^3 Eppendorf microcentrifuge tube and mix at 0 °C for 1 hour.
8. Transfer the slurry to a column made from a 1 cm^3 disposable syringe plugged with glass wool and allow to drain.
9. Wash the bed thoroughly with 5–10 cm^3 PBS containing 1% NP-40 (no BSA)[h].
10. Elute the bound immunoprecipitate into a 1.5 cm^3 Eppendorf tube with 3 successive washes of 1 M acetic acid (0.4 cm^3 each).
11. Add 70 μl 100% TCA and incubate at 0 °C for 30 min.
12. Recover the immunoprecipitate by spinning in a microcentrifuge for 2 min.
13. Remove supernatant, wash the pellet twice with acetone, and vacuum-desiccate[i].
14. Dissolve the pellet in 25 μl 0.1 M Na_2CO_3 containing 0.1 M dithiothreitol.
15. Mix with an equal volume of 4% SDS in 20% glycerol containing bromophenol blue as a tracking dye.
16. Heat at 80 °C for 5 min.
17. Remove aliquots of the dissolved immunoprecipitate to determine the incorporated radioactivity by liquid scintillation counting and analyse the remainder by SDS-PAGE.

IV. Immunoprecipitation of polysomes

Care must be taken during the isolation of polysomes in minimising the risk of degradation by the ubiquitous ribonucleases. This can best be achieved by the inclusion of ribonuclease inhibitors in the extraction buffers. Heparin, 2′+3′ AMP, detergent and a buffer at high pH and low temperature all help to reduce RNase activity. Trichodermin, a reported inhibitor of translation termination (Cundliffe, Cannon & Davies, 1974; Wei *et al.*, 1974) may help to increase polysome yields. All procedures are carried out at 0–4 °C using sterilized solutions and equipment.

1. To isolate polysomes, freeze tissue in liquid nitrogen, grind to a powder using a coffee grinder and mix well in a mortar with polysome extraction buffer – 200 mM Tris-HCl, pH 9.0; 120 mM KCl; 50 mM Mg Cl_2; 0.8 mg cm^{-3} dithiothreitol; 0.8 mg cm^{-3} 2′+3′ AMP; 0.1 mg cm^{-3} Na Heparin – 1 μg cm^{-3} trichodermin; 6.8% (w/w) sucrose[j].
2. Centrifuge in 100 cm^3 tubes at 150 × g for 10 min and decant the supernatant through 2 layers of sterile muslin.
3. Add Triton X-100 (a stock solution of 10% in polysome buffer) to a concentration of 1% and stir for 10 min.
4. Transfer to 40 cm^3 tubes and spin at 34,000 × g for 25 min.
5. Aspirate the supernatant and layer over 4 ml cushions of 54% (w/w) sucrose in polysome buffer in 30 cm^3 ultracentrifuge tubes.
6. Spin at 98,000 × g for 4 hours and decant the supernatant.
7. Rinse the polysome pellets with suspension buffer – 50 mM Tris-HCl, pH 7.5; 300 mM NaCl; 5 mM $MgCl_2$; 0.1% NP-40; 0.1 mg cm^{-3} Na Heparin; 1 μg cm^{-3} trichodermin.
8. Resuspend the polysomes in this buffer to a low concentration corresponding to A_{260} of 5–10 units cm^{-3}.
9. Remove unresuspended polysomes by centrifugation at 17,000 × g for 10 min[k].
10. Incubate the polysomes with the antibodies by gently mixing ribonuclease-free IgG with polysomes (approx. 25 μg per A_{260} unit) and incubating overnight at 0 °C.
11. Isolate the antibody-bound polysomes by passing the antibody-polysome solution twice through a column of Protein A Sepharose equilibrated in polysome suspension buffer at a slow flow rate of 20 cm^3 $hour^{-1}$.
12. Wash the column with 30 cm^3 buffer.
13. Dissociate and elute the bound polysomes with 15 cm^3 25 mM Tris-HCl, pH 7.5; 20 mM Na_2EDTA.
14. Isolate polyadenylated RNA by making the Sepharose-eluate 12 mM Tris-HCl, pH 7.5; 0.5 M NaCl; 10 mM Na_2EDTA; 0.2% SDS using concentrated solutions of the components.
15. Heat to 60 °C for 5 min.
16. Cool the solution to room temperature and pass it over a 0.4 cm^3 column of oligo-d(T) cellulose equilibrated with step 14 buffer at a flow rate of 15 cm^3 $hour^{-1}$.

17. Wash the column with this buffer until the A_{260} of the eluate returns to its baseline value.
18. Elute the mRNA with 0.8 cm^3 sterile distilled water[l].
19. Add 5 μg calf thymus tRNA (carrier), 80 μl 4M ammonium acetate, 2 cm^3 ethanol and precipitate at −20 °C overnight.
20. Recover the RNA by centrifuging for 15 min in a microcentrifuge.
21. Wash the pellet 3 times with 75% ethanol.
22. Vacuum desiccate the RNA preparation.
23. Dissolve in 10 μl sterile distilled water.
24. Check the purity of the preparation by translation, *in vitro*.

Notes

a For the screening of monoclonal antibodies the stacking gel is not divided into individual "slots" with a well-forming comb. Instead, a wide "slot" is formed extending over most of the width of the gel (see diagram in Fig. 3). For analysing multiple protein samples on a single gel, with an individual antibody, slots are formed to create different "tracks" as usual.

b Alternatively a "homemade" system can be contrived using two steel plates as electrodes. The polarity of the electric field should be such that the gel is nearer the negative pole and the membrane nearer the positive pole. For an SDS-containing gel of 20% polyacrylamide, a voltage of 50V typically results in a current of 200 mA. These conditions permit transfer from a 20% polyacrylamide gel in 3 hours.

c The BSA blocks all other non-specific protein binding sites, thus preventing spurious "background" binding of antibody.

d For screening hybridoma cultures, we have used supernatant from 1 cm^3 of cultured cells as the source of the antibody. For screening proteins transferred from polyacrylamide gels to filters 20 cm × 20 cm in area, as little as 10 μg IgG in 30 cm^3 PBS-BSA solution can be used for reaction with filter-bound antigens.

e Steps 8 and 9 should be performed on an orbital shaker in a high-efficiency fume hood in an area designated for such hazards.

f The presence of RNA in the immunoprecipitation reaction contributes to aggregation and digestion with RNAase reduces this.

g Boiling the reaction mixture in 1% SDS ensures that the components are all fully dissolved and reduces background markedly. BSA is added as a non-specific carrier, again to reduce background adsorption. It is important to remember that the immunoprecipitation is carried out in the presence of 0.1% SDS. It is probably advisable, therefore, to use an antiserum raised against an SDS-denatured antigen for this operation. Such antisera may not be suitable for experiments in which native proteins are probed with antibodies.

h Washing is most effective if the wash solution is introduced violently to the syringe-column, resuspending the bed. If reticulocyte lysates are being processed in this way, the washing of the column can be monitored by the disappearance of the pink colour (haemoglobin) from the bed.

i Acetone precipitates are sometimes difficult to dissolve in SDS-PAGE sample buffers. Step 14 is most efficacious.

j As a guide, for a preparation from *Phaseolus vulgaris* root nodules, 58 g tissue was homogenized in 120 cm^3 buffer.

k Steps 7–9 are important in preventing the non-specific trapping of aggregates in the Sepharose.

l It is important to exercise care at this stage, as the small amount of specific mRNA will probably be undetectable by an increase in absorbance at 260 nm.

KEY REFERENCES

Cullimore, J. V. & Miflin, B. J. (1983). Glutamine synthetase from the plant fraction of *Phaseolus* root nodules. Purification of the mRNAS and *in vitro* synthesis of the enzyme. *FEBS Letters*. **158**, 107–12.

Cuming, A. C. & Bennett, J. (1981). Biosynthesis of the light-harvesting chlorophyll a/b protein. Control of messenger RNA activity by light. *European Journal of Biochemistry*, **118**, 71–80.

Towbin, H., Staehelin, T. & Gordon, G. (1979). Electrophoretic transfer of proteins from polyacrylamide gels to nitrocellulose sheets: procedures and some applications. *Proceedings of the National Academy of Sciences, U.S.A.*, **76**, 4350–4.

Vaessen, R. T. M. J., Kreike, J. & Groot, G. (1981). Protein transfer to nitrocellulose filters. A simple method for the quantification of single proteins in complex mixtures. *FEBS Letters*, **124**, 193–6.

REFERENCES

Bedbrook, J. R., Smith, S. M. & Ellis, R. J. (1980). Molecular cloning and sequencing of cDNA encoding the precursor to the small subunit of chloroplast ribulose-1,5-bisphosphate carboxylase. *Nature*, **287**, 692–7.

Broome, S. & Gilbert, W. (1978). Immunological screening method to detect specific translation products. *Proceedings of the National Academy of Science, U.S.A.*, **75**, 2746–9.

Cowie, D. B., Spiegelman, S., Roberts, R. B. & Duerksen, J. D. (1961). Ribosome-bound beta-galactosidase. *Proceedings of the National Academy of Science, U.S.A.*, **47**, 114–22.

Cullimore, J. V. & Miflin, B. J. (1983). Glutamine synthetase from the plant fraction of *Phaseolus* root nodules. Purification of the mRNA and *in vitro* synthesis of the enzyme. *FEBS Letters*, **158**, 107–12.

Cullimore, J. V., Gebhardt, C., Saarelainen, R., Miflin, B. J., Idler, K. B. & Barker, R. F. (1985). Glutamine synthetase of *Phaseolus vulgaris* L.: organ-specific expression of a multigene family. *Journal of Molecular and Applied Genetics*, **2**, in press.

Cundliffe, E., Cannon, M. & Davies, J. (1974). Mechanism of inhibition of eukaryotic protein synthesis by trichothecene fungal toxins. *Proceedings of the National Academy of Science, U.S.A.*, **71**, 30–4.

Cuming, A. C. & Bennett, J. (1981). Biosynthesis of the light-harvesting chlorophyll a/b protein. Control of messenger RNA activity by light. *European Journal of Biochemistry*, **118**, 71–80.

Grunstein, D. & Hogness, D.S. (1975). Colony hybridization: a method for the isolation of cloned DNAs that contain a specific gene. *Proceedings of the National Academy of Science, U.S.A.*, **72**, 3961–5.

Hjelm, H., Hjelm, K. & Sjoquist, J. (1972). Protein A from *Staphylococcus aureus*, its isolation by affinity chromatography and its use as an immunoadsorbent for isolation of immunoglobulins. *FEBS Letters*, **28**, 73–6.

Jenkins, G. I., Hartley, M. R. & Bennett, J. (1983). Photoregulation of chloroplast development: transcriptional and posttranslational controls? *Philosophical Transactions of the Royal Society, London, B*, **303**, 419–31.

Kemp, D. J. & Cowman, A. F. (1981). Direct immunoassay for detecting *Escherichia coli* colonies that contain polypeptides encoded by cloned DNA segments. *Proceedings of the National Academy of Science*, USA, **78**, 4520–4.

Koehler, G. & Milstein, C. (1975). Continuous culture of fused cells secreting antibody of predefined specificity. *Nature*, **256**, 495–7.

Korman, A. J., Knudsen, P. J., Kaufman, J. F. & Strominger, J. L. (1982). cDNA clones for the heavy chain of HLA-DR antigens obtained after immunopurification of polysomes by monoclonal antibody. *Proceedings of the National Academy of Science, U.S.A.*, **79**, 1844–8.

Kraus, J. P. & Rosenberg, L. E. (1982). Purification of low-abundance messenger RNAs from rat liver by polysome immunoadsorption. *Proceedings of the National Academy of Science, U.S.A.*, **79**, 4015–9.

Laemmli, U. K. (1970). Cleavage of structural proteins during the assembly of the head of bacteriophage T4. *Nature*, **227**, 680–85.

Mullet, J. E., Burke, J. J. & Arntzen, C. J. (1980). Chlorophyll proteins of photosystem I. *Plant Physiology*, **65**, 823–7.

O'Farrell, P. H. (1975). High resolution two-dimensional electrophoresis of proteins. *Journal of Biological Chemistry*, **250**, 4007–21.

Palacios, R., Palmiter, R. D. & Schimke, R. T. (1972). Identification and isolation of ovalbumin-synthesizing polysomes. I. Specific binding of ^{125}I-anti-ovalbumin to polysomes. *Journal of Biological Chemistry*, **247**, 2316–21.

Palacios, R., Sullivan, D., Summers, N. M., Kiely M. L. & Schimke, R. T. (1973). Purification of ovalbumin-synthesizing polysomes and millipore partition of ribonucleic acid. *Journal of Biological Chemistry*, **248**, 540–8.

Palmiter, R. D., Palacios, R. & Schimke, R. T. (1972). Identification and isolation of ovalbumin-synthesizing polysomes. II. Quantification and immunoprecipitation of polysomes. *Journal of Biological Chemistry*, **247**, 3296–04.

Pelham, H. R. B. & Jackson, R. J. (1976). An efficient mRNA-dependent translation system from reticulocyte lysates. *European Journal of Biochemical*, **67**, 247–56.

Roberts, B. E. & Paterson, B. M. (1973). Efficient translation of tobacco mosaic virus RNA and rabbit globin 9S RNA in a cell-free system from commercial wheat germ. *Proceedings of the National Academy of Science, U.S.A.*, **70**, 2330–4.

Schechter, I. (1973). Biologically and chemically pure mRNA coding for a mouse immunoglobulin L-chain prepared with the aid of antibodies and immobilized oligothymidine. *Proceedings of the National Academy of Science, U.S.A.*, **70**, 2256–60.

Schimke, R. T., Palacios, R., Sullivan, D., Kiely, M. L., Gonzales, C. & Taylor, J. M. (1974). Immunoabsorption of ovalbumin synthesizing polysomes and partial purification of ovalbumin messenger RNA. *Methods in Enzymology*, **30**, 631–48.

Shapiro, S. Z. & Young, J. R. (1981). An immunochemical method for mRNA purification. Application to messenger RNA encoding trypanosome variable surface antigens. *Journal of Biological Chemistry*, **256**, 1495–8.

Southern, E. M. (1975). Detection of specific sequences among DNA fragments separated by gel electrophoresis. *Journal of Molecular Biology*, **98**, 503–17.

Towbin, H., Staehelin, T. & Gordon, G. (1979). Electrophoretic transfer of proteins from polyacrylamide gels to nitrocellulose sheets: procedures and some applications. *Proceedings of the National Academy of Science, U.S.A.*, **76**, 4350–4.

Vaessen, R. T. M. J., Kreike, J. & Groot, G. (1981). Protein transfer to nitrocellulose filters. A simple method for quantitation of single proteins in complex mixtures. *FEBS Letters*, **124**, 193–6.

Vodkin, L. O. (1981). Isolation and characterization of messenger RNAs for the seed lectin and Kunitz trypsin inhibitor of soybeans. *Plant Physiology*, **68**, 766–71.

Wei, C-M., Hansen, B. S., Vaughan, M. H., Jr. & McLaughlin, C. S. (1974). Mechanism of action of the mycotoxin trichodermin, a 12, 13-epoxytrichothecene. *Proceedings of the National Academy of Science, U.S.A.*, **71**, 713–7.

Young, R. A. & Davis, R. W. (1983). Efficient isolation of genes by using antibody probes. *Proceedings of the National Academy of Science, U.S.A.*, **80**, 1194–8.

N.J. BREWIN, G.W. BUTCHER, G. GALFRÈ, A.P. LARKINS, B. WELLS, E.A. WOOD & J.G. ROBERTSON

Immunochemical analysis of the legume root nodule

Introduction

Research into the *Rhizobium*-legume symbiosis has been flourishing for more than a century (Hellriegel, 1886; Fred, Baldwin & McCoy, 1932; Downie *et al.*, 1984). One reason for this is because of the economic importance of nitrogen fixation by legumes, which, on a global scale, accounts for almost half of the atmospheric nitrogen that is converted to ammonia and thereby introduced into the biological nitrogen cycle (Postgate, 1982). Another reason is because of the intrinsic academic interest in a biological system that involves the induction of an entirely new plant organ (the root nodule), a differentiated form of bacterium (the bacteroid) and a delicately controlled physiologial interaction between an invasive non-pathogenic prokaryote and the eukaryote partner (see Fig. 1–4).

As with any field of scientific endeavour, a major limitation to research progress has been the development of experimental techniques that enable us to pose and answer pertinent questions: the use of antibodies is almost certain to provide the next great leap forward for our study of the *Rhizobium*-legume symbiosis. To put the subject into historical perspective, there have been three streams of research activity in recent decades – cytology, biochemistry and genetics. The exciting prospect for immunochemical studies is that they should contribute to each of these three disciplines and result in a much more unified description of the molecular biology of the symbiosis.

Three kinds of antibody system have been used to investigate root nodules – polyspecific antibodies, monospecific antibodies and monoclonal antibodies (McAb). Polyspecific antisera have been raised against a plant cell supernatant fraction from homogenized root nodules of soybean (Legocki and Verma, 1980) and pea (Bisseling *et al.*, 1983). The antisera were then immunoprecipitated with a preparation from uninfected root tissue so that all the antibodies remaining in the solution were to antigens that are unique to the nodule (defined as “nodulins”). These nodulins could be identified after 2-dimensional gel electrophoresis, electroblotting

Fig. 1–4. Anatomy of a pea root nodule at increasing levels of magnification.

Fig. 1. Root nodules, *circa* 4 mm long.

Fig. 2. Longitudinal section through root nodule showing nodule meristem (nm), zone of bacteroid-containing cells (b), zone of senescence (s). The nitrogen-fixing activity of a nodule is a dynamic equilibrium between meristematic activity and bacterial infection on the one hand, and plant and bacteroid cell senescence on the other.

Fig. 3. High power light micrograph showing several thousand nitrogen-fixing bacteroids (Ba) within the cytoplasm of a single plant cell. Key: nucleus – n; cell wall – cw; starch granules – sg.

Fig. 4. Electron micrograph showing bacteroids (Ba) surrounded by peribacteroid membrane (pbm) of plant origin. Nodule bacteria are released individually into the plant cell cytoplasm from an infection thread which is bound by a membrane that is continuous with the plant cell plasma membrane.

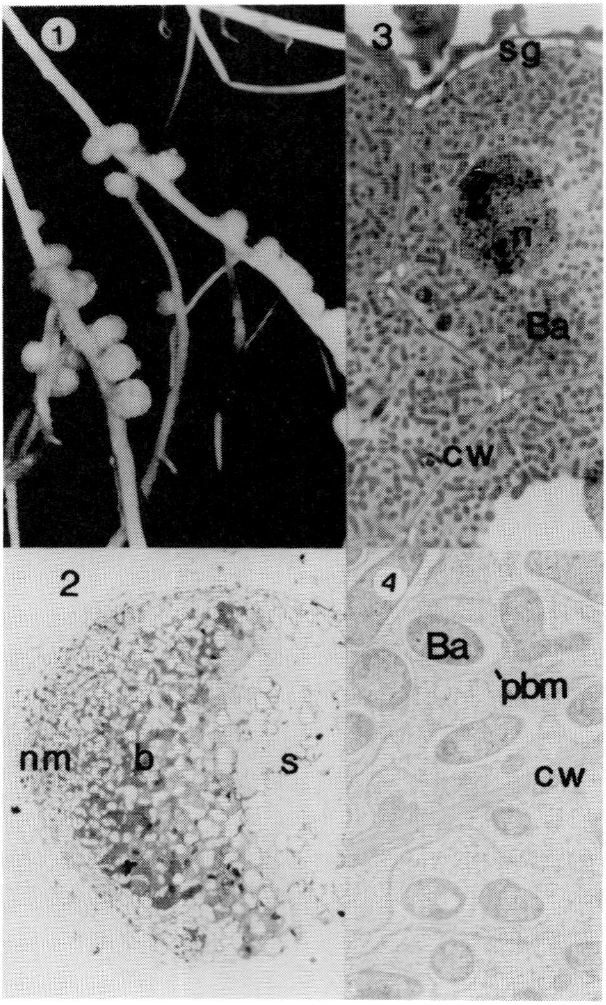

to nitrocellulose sheets (see chapter by Cuming *et al.*, this volume, for details of the methods) and probing with the nodule-specific antiserum coupled to iodine-125. About 40 nodule-specific polypeptides were identified in this way and these have been subdivided into those that arise early and those that arise late during nodule development. The biological function of one nodulin has been deduced – nodulin-35 has been identified as uricase, an enzyme involved in the assimilation of fixed nitrogen within soybean nodules. A monospecific antibody has been prepared for nodulin-35 and used for immunofluorescence staining of nodule tissue sections; the enzyme appears to be localized to the uninfected interstitial cells of the nodule tissue (Bergmann, Preddie & Verma, 1983).

Monospecific (serum) antibodies have also been obtained to other purified components of the root nodule. Antiserum to leghaemoglobin (the most abundant protein in the root nodule) has been used to obtain an enriched messenger RNA fraction by immunoprecipitation of polysomes. This, in turn, was used to obtain cDNA clones and hence genomic clones which have now been sequenced (Sullivan *et al.*, 1981; Wiborg *et al.*, 1983). Similarly a monospecific antibody to purified glutamine synthetase has been isolated (Cullimore *et al.*, 1983) and used to obtain cDNA and genomic clones for this gene.

The remainder of this article will be concerned with a particular immunochemical research programme that we have initiated which is designed to investigate the surface interactions between the endosymbiotic nitrogen-fixing bacteroid and the plant-derived peribacteroid membrane system that envelopes each bacteroid individually (Fig. 5). This system

Fig. 5. Analysis of surface components of the peribacteroid membrane, peribacteroid space and bacteroid surface using monoclonal antibodies for immunocytochemistry. The symbols represent possible antigens which could be identified and analysed using monoclonal antibodies.

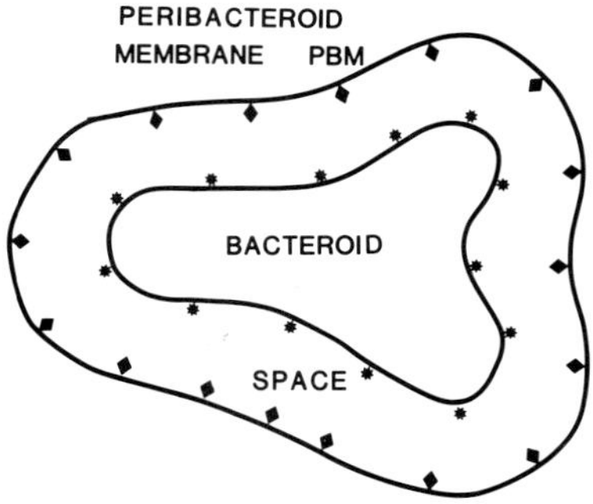

merits detailed investigation firstly, because it must control the exchange of any metabolite between the cytoplasm of the plant cell and the bacteroid and, secondly, because it is the interface for any surface recognition signals between the bacterium and its host that may serve to maintain normal nodule morphogenesis and prevent the bacterium from being treated as an invading pathogen.

The technical problems inherent in this research are that the relevant antigens are all completely uncharacterized and that very small amounts of material are involved. Moreover, the most interesting antigens are likely to be those that are unique to the nodule state. Because of the difficulties in obtaining purified antigens, the best strategy seemed to be to generate a wide range of McAb that reacted with the peribacteroid membrane, the bacteroid surface and the peribacteroid space (Fig. 5). From among these

Fig. 6. Simplified scheme for the isolation of membrane-enclosed bacteroids from homogenized pea root nodules in the presence of 0.5 M sucrose – see Brewin *et al.*, 1985 for details.

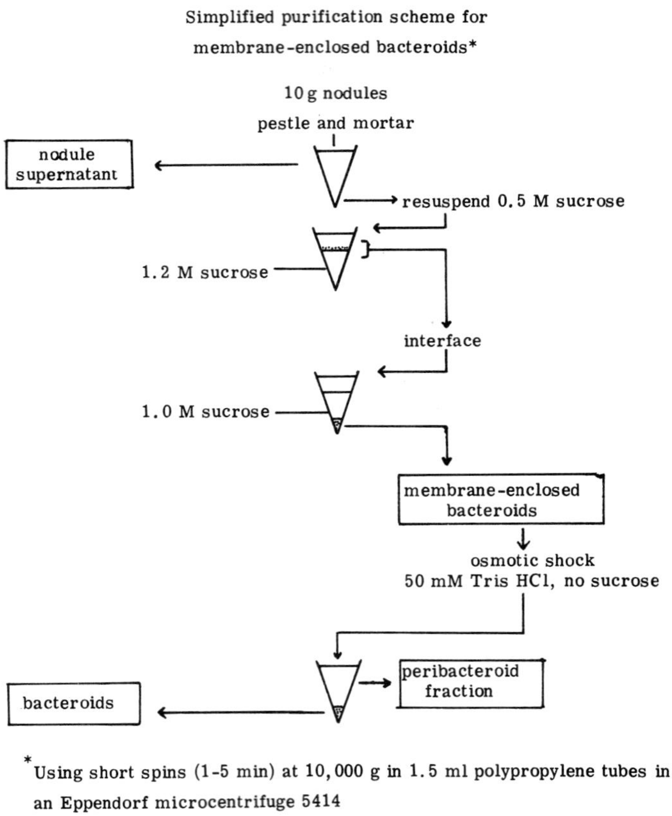

McAb, those that reacted to interesting antigens could be selected for further study using the techniques of cytology, biochemistry and molecular genetics.

What makes this immunochemical approach feasible is the availability of a rapid and simple technique for isolating membrane-enclosed bacteroids by differential centrifugation of pea root nodules homogenized in the presence of 0.5 M sucrose (Fig. 6). The enclosing peribacteroid membrane was ruptured by a mild osmotic shock treatment (without sucrose) so that a subsequent centrifugation step separated the bacteroid cells from a supernatant fraction containing peribacteroid membrane and the contents of the peribacteroid space (Fig. 7). These fractions, either the sonicated bacteroids or the peribacteroid material, were used as antigens for the immunization of rats.

Fig. 7. Examination of fractions used as antigens by SDS polyacrylamide gel electrophoresis. A – after staining with Coomassie blue; B – after periodate-silver staining for carbohydrates and glycoproteins; C – after electroblotting to nitrocellulose sheets, probing with monoclonal antibody AFRC MAC 57 (to bacterial LPS) followed by immunoperoxidase staining; D – after electroblotting and treatment with AFRC MAC 64, a monoclonal reacting to the peribacteroid membrane. Fractions: 1 – free-living *Rhizobium leguminosarum* 3841; 2 – sonicated bacteroid supernatant; 3 – peribacteroid material.

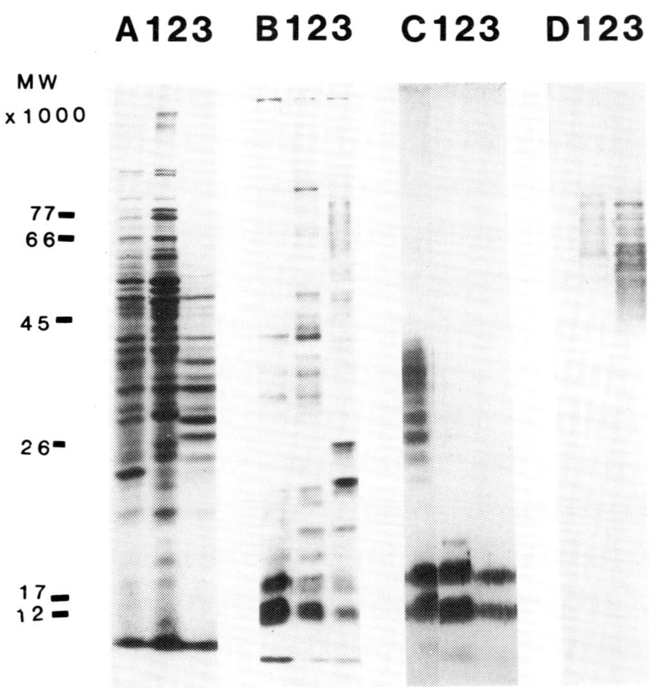

Monoclonal antibody for bacteroid lipopolysaccharide

In the case of the bacteroid material, a sonicated cell supernatant fraction was used as antigen. After a series of immunizations an antiserum was obtained which at 1:1000 dilution still reacted with antigen in the dot immunoassay (see Roberts, this volume, for details of the dot immunobinding method). Spleen cells from the rat were then fused with a myeloma line and hybrid cell lines were selected (Galfrè and Milstein, 1981). Cell supernatants from these hybrid lines were then tested for antibody activity against nanogram quantities of the original antigen dotted onto nitrocellulose sheets. Antibody binding was detected by the indirect immunoperoxidase technique (Smith *et al.*, 1984), using peroxidase coupled to rabbit anti-rat immunoglobulin and then staining with 1,4-chloronaphthol and hydrogen peroxide. Positive McAb lines were then screened in various ways in order to pick up as wide a range of McAb as possible. They were tested by the dot immunobinding assay (Fig. 8) against bacteroid preparations from other strains of *Rhizobium leguminosarum*, in order to identify strain-specific antigens, and against free living cultures of the strain used as antigen (strain 3841, Wang, Wood & Brewin, 1982) in order to identify bacteroid-specific antigens. Subsequently, the antigen for each McAb was identified after SDS polyacrylamide gel electrophoresis (SDS PAGE), electroblotting (Towbin, Stachelin & Gordon, 1979) and staining by the indirect immunoperoxidase technique described above.

After five separate fusion experiments involving three different rats, the picture that emerged was as follows. More than 90% of the positive cell lines produced antibodies to bacteroid lipopolysaccharide (LPS). None of these antibodies was to an antigen that was specific for the bacteroid state (although some of the antigens were strain specific and LPS from different strains of *R. leguminosarum* had different mobilities on SDS PAGE gels). However, what was noticeable by examination of electroblots was that the LPS from bacteroids of *R. leguminosarum* strain 3841 was in a low molecular weight form (presumably equivalent to the 'rough' LPS of *Salmonella* and *E. coli*) whereas in free-living cultures of the same strain about half the LPS was of higher molecular weight, presumably as a result of substitution with O-antigen side chains to give the 'smooth' form of LPS (Carlson, 1984). Thus it appears that the bacteroid cells lack O-antigen side chains on their LPS and it is possible that this helps them to avoid eliciting the host defence reactions of the pea. It was also noted that significant amounts of the low molecular weight LPS were present in the peribacteroid material released from membrane-enclosed bacteroids by mild osmotic shock treatment. This result could indicate that there is a very low concentration of divalent cations in the peribacteroid space since, in free-living bacteria, treatment with EDTA is a standard method for releasing LPS from the cell wall (Leive and Morrison, 1975).

The presence of highly antigenic LPS in the peribacteroid material represented an obstacle to obtaining a wide range of McAb to other antigens of the peribacteroid membrane and peribacteroid space. For this reason a McAb reacting with LPS, AFRC MAC 57, was purified and coupled to activated CH Sepharose 4B. This material was then used as an affinity column to remove LPS from the peribacteroid material before immunizing the rats. By the same method it should also be possible to remove LPS from the sonicated bacteroid preparation so that a wide range of McAb can more readily be obtained. We are particularly interested in cell surface antigens that are more abundant in the bacteroid than in the free-living state. So far, from about 200 McAb that reacted with bacteroid antigens, only one, MAC 90, seems not to react with free living bacteria. The properties of this particular antigen are currently being investigated.

Tissue localization of antigens with immunogold

In addition to studying the bacteroid cell surface, we also want to use McAb to identify antigens on the plant-derived peribacteroid membrane and in the peribacteroid space. Our long-term hope is that the study of these antigens will teach us about transport processes between the bacteroid and plant cytoplasms, about plant membrane biogenesis and recycling, and about surface interactions between an invasive 'non-pathogenic' bacterium and its eukaryotic host. How could we be sure that our McAb was reacting with a component of the peribacteroid membrane or peribacteroid space? Although rapid and simple, the tissue fractionation procedure described in Fig. 6 is far from complete and gives considerable cross-contamination. Therefore, it was necessary to examine the ultrastructural localization of antigens in thin sections of nodule tissue embedded for electron microscopy. In order to achieve this, the normal procedures for tissue fixation and embedding had to be adapted so that antigenicity was preserved (Protocol I).

Sliced pea nodules were fixed in glutaraldehyde and dehydrated through an alcohol series at 0 °C, −20 °C and −35 °C. They were then infiltrated with Lowicryl K4M resin which was polymerized by UV light, first at −40 °C for 24 h and then at room temperature (Robertson *et al.*, 1984, 1985). Without osmium treatment, the tissue became almost transparent during preparation and therefore the plant material was held between two pieces of nylon mesh within a simple tissue-handling device during the low temperature procedures, which were carried out in a box cooled with liquid nitrogen (Wells, 1985). At the end of these manipulations, the nodule slices still had the characteristic pink colour of leghaemoglobin, suggesting that protein structure had not been severely denatured. This indication that antigenicity had been maintained was confirmed when thin sections of tissue were transferred to gold grids and treated with rabbit antiserum to purified leghaemoglobin, followed by goat anti-rabbit IgG conjugated to colloidal gold. It was clearly shown that leghaemoglobin was localized to the cytoplasm of infected plant cells (Robertson *et al.*, 1984 and Fig. 9*A*). Similarly, treatment of nodule thin sections with the McAb, MAC 57 (which reacts with bacteroid LPS), resulted in gold staining of the bacteroid cell wall (Fig. 9*B*).

Identification of antigens in the peribacteroid membrane

Antibody-producing myeloma cell lines were isolated from two rats that had been immunized with peribacteroid material (i.e. the material released from membrane-enclosed bacteroids by mild osmotic shock treatment). Approximately 100 cell lines were screened by dot immunoassay and three lines were identified that produced McAb which reacted with peribacteroid material but did not react with antigens from sonicated

Fig. 8. Screening of monoclonal antibodies by dot immunoassay and immunoperoxidase staining of nitrocellulose sheets. Antigens (horizontal rows): 1 – peribacteroid fraction; 2 – strain 3841 bacteroid; 3 – strain 3841 (free-living culture); 4 – blank (1% BSA); 5 – strain B155 bacteroid. Monoclonal antibody cell supernatants: A – AFRC MAC 64 (to peribacteroid membrane); B – AFRC MAC 57 (to lipopolysaccharide); C – AFRC MAC 90 (to bacteroid-specific cell wall component).

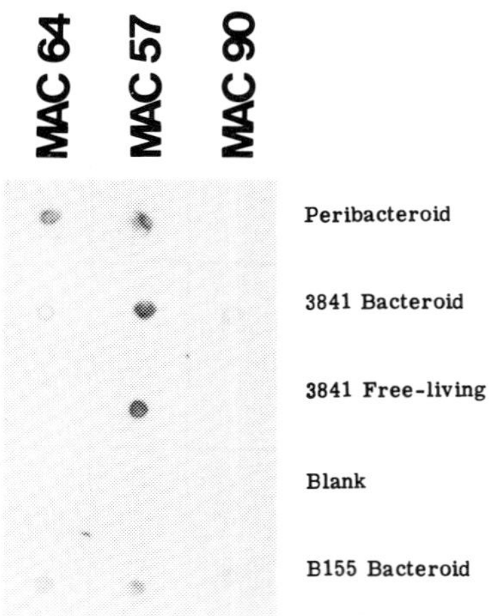

free-living rhizobia (see Fig. 8, AFRC MAC 64). A further 10 McAb reacted with antigens from peribacteroid, bacteroid and free-living bacteroid samples – these monoclonal antibodies have not been examined in detail, but the majority were to bacterial LPS.

The three McAb which in dot immunoassay reacted strongly with the peribacteroid fraction were used to identify the intracellular location of the corresponding antigens by immunogold labelling. In each case a membrane antigen was involved; the antigen was most abundant in the peribacteroid membrane (Fig. 9*C*) but gold-labelling of the plant cell plasma membrane, the bounding membrane of infection threads and the Golgi apparatus was also observed (Brewin *et al.*, 1985). However, the endoplasmic reticulum, nuclear membrane and other plant membranes were not stained. (In this and all subsequent tests described below the three monoclonal antibodies behaved identically, indicating that this particular antigen had been strongly selected for by our screening system.) The presence of a common antigen in the peribacteroid membrane, plasma membrane and Golgi systems is the first cytochemical evidence to support the hypothesis (previously based on cytological evidence) that the peribacteroid membrane and plasma membranes are each formed by fusion of Golgi-derived membrane vesicles (Brewin *et al.*, 1985).

This is one of the first reports of a McAb to an antigen in a plant cell membrane, and it was of interest to identify the biochemical nature of the antigen. Peribacteroid material was treated with 0.5% Triton X100 (to solubilize membrane material), fractionated by SDS PAGE and electroblotted to nitrocellulose sheets. These were then treated with the monoclonal antibody, AFRC MAC 64, and binding was detected by immunoperoxidase staining (Fig. 10). A range of antigen bands was observed corresponding to polypeptide molecular weights from 55K to 80K. Pretreatment of the peribacteroid material with 1 mg cm^{-3} pronase for 1 h at 30 °C completely eliminated all antibody binding after gel electrophoresis and electroblotting, indicating a polypeptide component in the antigen. Pre-treatment with trypsin (1 mg cm^{-3}) or lower concentrations of pronase resulted in loss of some, but not all, antigen bands indicating that the polypeptides were somewhat resistant to proteolytic attack. Treatment of nitrocellulose blots with sodium periodate completely destroyed their ability to bind MAC 64 antibody, indicating that the antigen (or epitope) probably involved a carbohydrate, i.e. that the antibody was reacting with the carbohydrate component of a membrane glycoprotein or group of glycoproteins. Periodate and pronase treatments were also performed on thin sections of nodules prior to immunogold labelling – the periodate treatment again prevented binding of MAC 64 to peribacteroid membrane,

Fig. 9. Immunogold staining of thin sections of nodule tissue, fixed and embedded in Lowicryl K4M resin for electron microscopy. A – localization of leghaemoglobin to the cytoplasm of infected cells (pc) using antiserum to purified leghaemoglobin; B – localization of lipopolysaccharide in the bacteroid outer membrane (bom) using monoclonal antibody AFRC MAC 57; C – identification of the antigen for monoclonal antibody AFRC MAC 64 in the peribacteroid membrane (pbm) of infected root nodule cells. Key: bc – bacteroid cytoplasm; bom – bacteroid outer membrane; pbm – peribacteroid membrane; pc – plant cytoplasm; μ-bar represents 0.2 μM.

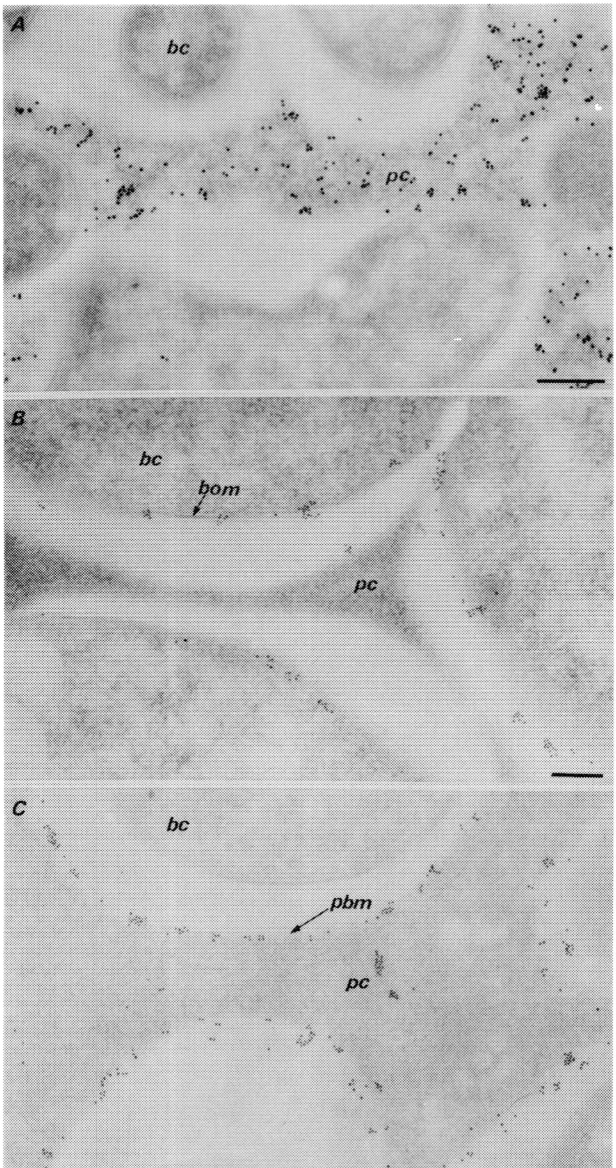

but pronase treatment did not inhibit antibody binding to thin sections. These results indicate that the epitope for MAC 64 consists solely of a carbohydrate side chain of the glycoprotein, perhaps without the involvement of a pronase-sensitive polypeptide component.

The reason for the dispersed series of antigen bands seen after SDS PAGE and electroblotting is not clear. Either it represents progressive

Fig. 10. Identification of AFRC MAC 64 antigen following fractionation of peribacteroid material by SDS PAGE, electroblotting to nitrocellulose and successive treatments with MAC 64, anti-rat IgG-peroxidase conjugate, 1,4-chloronaphthol and hydrogen peroxide. 1 – peribacteroid material; 2 – pretreated with 1 mg cm^{-3} trypsin; 3 – pretreatment with 1 mg cm^{-3} pronase. Note that periodate oxidation of nitrocellulose sheets prior to immuno-staining completely destroyed the ability of MAC 64 to bind.

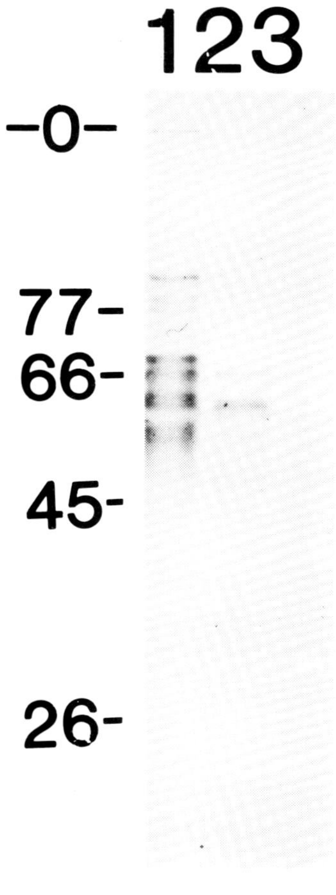

degradation or progressive carbohydrate substitution on a single glycoprotein species, or, the oligosaccharide recognized by MAC 64 may be added to a number of different membrane glycoproteins within the Golgi apparatus. It is also not clear why this particular epitope has generated so many apparently identical McAb. Another three similar McAb were isolated from a subsequent fusion experiment and, yet, the glycoprotein bands stained by MAC 64 after SDS PAGE correspond to relatively minor macromolecular components of the peribacteroid material as judged by protein or carbohydrate staining systems (Coomassie blue or periodate/silver staining in Fig. 7). Conceivably, the antigen exists in another molecular form which is highly immunogenic but which does not itself bind to nitrocellulose.

In order to obtain McAb to other components of the peribacteroid membrane and peribacteroid space it will first be necessary to construct an immunoaffinity column containing MAC 64 IgM so that this prevalent antigen can be removed from the peribacteroid material before immunization. The affinity column could also be used to purify the MAC 64 antigen for further biochemical analysis. Although the MAC 64 antigen is not specific to pea nodules (it was also found to be a component of Golgi membranes from onion meristems), it may still be an important component of the peribacteroid membrane simply by virtue of its abundance. In addition, the McAb could be used in studies of membrane biogenesis in plant cells.

Conclusions

Monoclonal antibodies are being used to identify and isolate the molecular components involved in the surface interactions between the endosymbiotic nitrogen-fixing bacteroid and the enveloping peribacteroid membrane of plant origin. Starting with relatively small quantities of uncharacterized and unpurified biological material, it has been possible to isolate monoclonal antibodies which can subsequently be used as specific molecular probes in a variety of ways: for immunocytology at light and electron microscope levels; for immunopurification and for biochemical analysis of the antigen; for genetical studies using antibody binding as the assay for gene products. So far, the most abundant classes of McAb to emerge from our screening programmes have been to the core fraction of bacteroid LPS and to a glycoprotein component of the peribacteroid membrane. As our screening systems become modified, a wider range of monoclonal antibodies to nodule-specific antigens should emerge and these will be very interesting candidates for further study, particularly in relation to the use of *Rhizobium* symbiotic mutants as experimental probes to discover more about plant morphogenesis and the control of plant pathogenicity.

PROTOCOLS

I. Fixation and embedding of legume root nodules in Lowicryl K4M

The use of immunochemistry on thin resin sections requires material that has retained its antigenicity and ultrastructure. This can be achieved by fixing the tissue with glutaraldehyde (step 1) and carrying out the dehydration (steps 2–7), infiltration (steps 8–13) and polymerization (step 14) at low temperatures (−20 to −35/40 °C). The handling of plant material under the above conditions can become an easy routine procedure when the low temperature box and tissue handling device are used as described in the text. Lowicryl K4M resin was obtained from TAAB Laboratories, Reading, UK.

1. Fix tissue for 2 h at room temperature in 2.5% glutaraldehyde, 0.05 M sodium cacodylate pH 7.2[a].
2. Dehydrate in 30% EtOH 1 h on ice;
3. 50% EtOH 1 h at −20 °C[b];
4. 70% EtOH 1 h at −35 °C;
5. 95% EtOH 1 h at −35 °C;
6. 100% EtOH 1 h at −35 °C;
7. 100% EtOH 1 h at −35 °C.
8. Infiltrate in resin:ethanol (1:1) for 1 h at −35 °C;
9. 2:1 resin:ethanol 1 h at −35 °C;
10. 3:1 resin:ethanol 1 h at −35 °C;
11. 100% resin 1 h at −35 °C;
12. 100% resin overnight at −35 °C;
13. 100% resin 8 h at −35 °C.
14. Polymerize for 24 h with indirect UV irradiation at −40 °C (light source: EYE FL20S BLB);
15. Polymerize for 48 h with indirect UV irradiation at room temperature.

II. Immunogold labelling of thin sections embedded for electron microscopy in Lowicryl K4M

Immunogold labelling of thin sections is carried out under conditions that prevent non-specific absorption of the antibodies. The use of the indirect immunolabelling method makes possible the use of commercially available gold conjugated secondary antibodies.

1. Cut sections with 45 ° glass knife, fill trough with water, then lower the level, if the block face wets, blot it and cut a few thick sections, then continue cutting thin sections.
2. Pick up sections onto 200 mesh gold grids that have a carbon plastic film, and have been glow-discharged in the presence of amines.
3. Submerge grids in 20 mg cm^{-3} bovine serum albumin (BSA) in TBS/P[c] at 37 °C for 1 h.
4. Transfer grids to primary antibody (e.g. rabbit anti-leghaemogloblin

antiserum), diluted 1:10 with 20 mg cm^{-3} BSA TBS/P at 37 °C for 1 h.

5. Wash grids with 2 mg cm^{-3} BSA in TBS/P, six times during 20 min, with agitation.
6. Submerge grids in gold labelled secondary antibody[d], diluted 1:10 with 2 mg cm^{-3} BSA TBS/P for 1 h at 37 °C.
7. Wash in 2 mg cm^{-3} BSA in TBS/P + 0.1% Triton X100, at least 3 times.
8. Wash 3 times in distilled water.
9. Stain on drops of saturated uranyl acetate for 10 min.
10. Stain on drops of lead citrate for 2 min.

Notes

a Tissue blocks were about 1 mm cube. When tissue floated, the trapped air was removed with a slight vacuum and the fixative replaced.

b At all stages after 30% ethanol, specimens were in a rotating block within a low temperature box (Wells, 1985). A version of this low temperature box can be obtained from Thames Cryogenics Ltd., Abingdon, Oxon.

c This blocks any sites on the tissue section which might bind antibody non-specifically. TBS/P 10 mM Tris HCl pH 7.4, 0.02% Na azide, 0.9% NaCl and 0.5 mg cm^{-3} Polyethylene glycol (20,000).

d Janssen Pharmaceuticals, e.g. Goat anti-rabbit IgG, coupled to 20 nm colloidal gold (or see chapter by Perrot-Rechenmann and Gadal, this volume, for the preparation of gold labels).

KEY REFERENCES

Brewin, N. J., Robertson, J. G., Wood, E. A., Wells, B., Larkins, A. P., Galfre, G. & Butcher, G.W. (1985). Monoclonal antibodies to antigens in the peribacteroid membrane from *Rhizobium*-induced root nodules of pea cross-react with plasma membranes and Golgi bodies. *EMBO J.*, **4**, 605–11.

Downie, J. A., Ma, Q. S., Knight, C. D., Rossen, L., Hombrecher, G., Brewin, N. J. & Johnston, A. W. B. (1984). Molecular Genetics of *Rhizobium leguminosarum* nodulation and nitrogen fixation. Annual Proceedings of the Phytochemical Society of Europe, vol. 22. In *The genetic manipulation of plants and its application to agriculture*, ed. P. J. Lea & G. R. Stewart, pp. 61–87. Oxford: Oxford University Press.

Legocki, R. P. & Verma, D. P. S. (1980). Identification of nodule-specific host proteins (nodulins) involved in the development of the *Rhizobium*-legume symbiosis. *Cell*, **20**, 153–63.

Robertson, J. G., Wells, B., Bisseling, T., Farnden, K. J. F. & Johnston, A. W. B. (1984). Immunogold localization of leghaemoglobin in cytoplasm in nitrogen-fixing root nodules of pea. *Nature*, **311**, 254–6.

Robertson, J. G., Wells, B., Brewin, N. J. & Williams, M. A. (1985). Immunogold localization of cellular constituents in the *Rhizobium*-legume symbiosis. In *Oxford Reviews of Plant Molecular and Cell Biology*, vol. 2, ed. B. J. Miflin, in the press. Oxford: Oxford University Press.

Sullivan, D., Brisson, N., Goodchild, B., Verma, D. P. S. & Thomas, D. (1981). Molecular cloning and organization of two leghaemoglobin genomic sequences of soybean. *Nature*, **289**, 516–8.

REFERENCES

Bergmann, H., Preddie, E. & Verma, D. P. S. (1983). Nodulin 35: a subunit of specific uricase (uricase II) induced and localized in the uninfected cells of soybean nodules. *EMBO Journal*, **2**, 2333–9.

Bisseling, T., Benn, C., Klugkist, J., van Kammen, A. & Nadler, K. (1983). Nodule-specific host proteins in effective and ineffective root nodules of *Pisum sativum*. *EMBO Journal*, **2**, 961–6.

Brewin, N. J., Robertson, J. G., Wood, E. A., Wells, B., Larkins, A. P., Galfrè, G. & Butcher, G. W. (1985). Monoclonal antibodies to antigens in the peribacteroid membrane from *Rhizobium*-induced root nodules of pea cross-react with plasma membranes and Golgi bodies. *EMBO Journal*, **4**, 605–11.

Carlson, R. W. (1984). Heterogeneity of *Rhizobium* lipopolysaccharide. *Journal of Bacteriology*, **158**, 1012–7.

Cullimore, J. V., Lara, M., Lea, P. J. & Miflin, B. J. (1983). Purification and properties of two forms of glutamine synthetase from the plant fraction of *Phaseolus* root nodules. *Planta*, **157**, 245–53.

Downie, J. A., Ma, Q. S., Knight, C. D., Rossen, L., Hombrecher, G., Brewin, N. J. & Johnston, A. W. B. (1984). Molecular genetics of *Rhizobium leguminosarum* nodulation and nitrogen fixation. Annual Proceedings of the Phytochemical Society of Europe, Vol. 22. In *The genetic manipulation of plants and its application to agriculture*, ed. P.J. Lea & G.R. Stewart, pp. 61–87. Oxford: Oxford University Press.

Fred, E. B., Baldwin, I. L. & McCoy, E. (1932). *University of Wisconsin studies in plant science, No. 5 Root nodule bacteria and leguminous plants*. Madison: University of Wisconsin.

Galfrè, G. & Milstein, C. (1981). Preparation of monoclonal antibodies: strategies and procedures. *Methods in Enzymology*, **73**, 3–46.

Hellriegel, H. (1886). *Z. Ver. Rubenzucker-Ind. Dt. Reichs*, **36**, 863–7.

Legocki, R. P. & Verma, D. P. S. (1980). Identification of nodule-specific host proteins (nodulins) involved in the development of the *Rhizobium*-legume symbiosis. *Cell*, **20**, 153–63.

Leive, L. & Morrison, D. C. (1975). Isolation of lipopolysaccharide from bacteria. *Methods in Enzymology*, **23**, 254–64.

Postgate, J. R. (1982). Fundamental principles of nitrogen fixation (C.U.P.).

Robertson, J. G., Wells, B., Bisseling, T., Farnden, K. J. F. & Johnston, A. W. B. (1984). Immunogold localization of leghaemoglobin in cytoplasm in nitrogen-fixing root nodules of pea. *Nature*, **311**, 254–6.

Robertson, J. G., Wells, B., Brewin, N. J. & Williams, M. A. (1985). Immunogold localisation of cellular constituents in the legume-*Rhizobium* symbiosis. In *Oxford Reviews of Plant Molecular and Cell Biology*, vol. **2**, ed. B. J. Miflin, pp. 69–89. Oxford: Oxford University Press.

Smith, E., Roberts, K., Butcher, G. W. & Galfre, G. (1984). Monoclonal antibody screening: two methods using antigens immobilized on nitrocellulose. *Analytical Biochemistry*, **138**, 119–24.

Sullivan, D., Brisson, N., Goodchild, B., Verma, D. P. S. & Thomas, D. (1981). Molecular cloning and organization of two leghaemoglobin genomic sequences of soybean. *Nature*, **289**, 516–18.

Towbin, H., Stachelin, T. & Gordon, J. (1979). Electrophoretic transfer of proteins from polyacrylamide gels to nitrocellulose sheets. *Proceedings of the National Academy of Sciences, U.S.A.*, **76**, 4350–4.

Wang, T. L., Wood, E. A. & Brewin, J. J. (1982). Growth regulators, *Rhizobium* and nodulation in peas. *Planta*, **155**, 345–9.

Wells, B. (1985). Low temperature box and tissue handling device for embedding biological tissue for immunostaining in electron microscopy. *Micron and Microscopie Acta*, **16**, 49–53.

Wiborg, O., Hyldig-Nielsen, J. J., Jensen, E., Paludan, K. & Marcker, K. A. (1983). Structure of an unusual leghaemoglobin from soybean. *EMBO Journal*, **2**, 449–52.

B. THOMAS, M.D. PARTIS & B.R. JORDAN

Immunological approaches to the study of phytochrome

Introduction

Phytochrome is a chromoprotein photoreceptor responsible for controlling the majority of developmental responses that plants show to light including seed germination, extension growth, de-etiolation changes and floral induction. It exists in two interconvertible forms; a red light-absorbing form called Pr and a far-red-absorbing form called Pfr which is believed to be the physiologically active configuration. The application of immunology has proved to be a particularly appropriate tool for studying phytochrome, as it is a large protein to which antibodies are readily formed. Immunology provides not only a means of detecting and quantifying phytochrome, but also offers a means of probing the molecular characteristics of phytochrome which are related to its mechanism of action. The aim of this chapter is to review how immunological techniques have been used in the study of this photoreceptor. Immunology is of course only one of a number of disciplines which have been exploited for the characterization of the phytochrome system. For this reason, findings based on immunological techniques have been placed in the context of results using complementary experimental approaches. The reader is also directed to a number of excellent review articles on these wider aspects of the subject (Marmé, 1977; Pratt, 1979; Quail *et al.*, 1983; Smith, 1983).

Molecular characteristics of phytochrome

The properties of antibodies in any preparation are inevitably determined by the antigens which elicit their formation. In this context it is useful to appreciate that many biochemical and immunological studies have been carried out using phytochrome differing from that which is presently believed to be the native form of the molecule. Phytochrome was initially purified from etiolated oat (*Avena sativa*) seedlings as a protein of molecular weight 60,000 (Mumford & Jenner, 1966), but it was soon appreciated from the disparity between the molecular weights of oat and rye (*Secale cereale*) phytochrome (Correll *et al.*, 1968), that the 60 kDa peptide represented a protease-resistant core of the native phytochrome molecule. Faster purification techniques resulted in the preparation of a 118–120 kDa protein which was called 'large' phytochrome to distinguish it from the 'small' chromophore-containing 60 kDa fragment. Small phytochrome was shown to arise through attack of endogenous proteases on large phytochrome during purification (Gardner *et al.*, 1971) and this was confirmed using antisera to either large or small phytochrome (Cundiff & Pratt, 1973).

For the next ten years or so, most immunological and biochemical investigations used large phytochrome in the belief that it represented an undegraded form of the molecule. More recent studies provided evidence that this assumption was in error and that large phytochrome was itself a mixture of degradation products of the native molecule. Large phytochrome prepared by conventional methods (Rice & Briggs, 1973*a*) or by immunoaffinity chromatography (Hunt & Pratt, 1980*a*) possessed multiple N-termini and also showed multiple bands in immunodiffusion (Rice & Briggs, 1973*b*) or in isoelectric focussing (Vierstra & Quail, 1982*a*). Furthermore, widely differing amino acid compositions were obtained for phytochrome prepared by different methods but from the same source. Boeshore & Pratt (1981) found that following immunoprecipitation, phytochrome from red-light irradiated oat seedlings exhibited reduced electrophoretic mobility compared to that from far-red light irradiated tissue and concluded that the molecule was modified in some way on photoconversion to Pfr. An alternative proposal was advanced by Vierstra & Quail (1982*a*) using anti-phytochrome coated *Staphylococcus aureus* cells to absorb phytochrome from extracts of etiolated oat seedlings. Extraction in hot SDS-containing buffers yielded phytochrome of molecular weight 124 kDa irrespective of the spectral form in which the phytochrome was extracted (Vierstra & Quail, 1982*a*; Bolton & Quail, 1982), while the use of detergent-free buffers produced phytochrome of 124 kDa only when extracted as Pfr; extraction as Pr resulted in a heterogeneous mixture of 114–118 kDa peptides. They further showed that incubation of crude extracts as Pr resulted in the progressive conversion of the 124 kDa phytochrome to the 114/118 kDa peptides and that the conversion could be prevented by the protease inhibitor PMSF (Vierstra & Quail, 1982*a*). Kerscher and Nowitzki (1982) extracted phytochrome as Pr or Pfr from etiolated rye seedlings and determined the molecular weights of the peptide species present by electrophoresis, Western blotting and immunodetection with peroxidase-labelled anti-phytochrome IgG. They also found that SDS-extracted phytochrome had a molecular weight of 124 kDa and that degradation to a mixture of 114/118 kDa proteins was observed when detergent-free extracts were incubated as Pr. A similar proteolytic degradation of the native phytochrome molecule has been shown to occur in other plant species (Kerscher & Nowitzki, 1982; Vierstra *et al.*, 1984) by Western blotting. These studies, along with the observations that the *in vitro* translation product of mRNA for the phytochrome apoprotein also has a molecular weight of 124 kDa (see below), strongly suggest that this represents the size of the native molecule. In this chapter we call this 'undegraded' phytochrome to distinguish it from 'small' and 'large' phytochrome mentioned earlier. Although the change in the molecular weight is small, the properties of phytochrome in the large form (114/118 kDa) differed substantially from those of the 124 kDa undegraded protein (Quail *et al.*, 1983), and have promoted a re-evaluation of hypotheses of the mode of action of phytochrome which were based on the properties of large phytochrome.

Phytochrome assays

Potentially, one of the most important applications of immunochemistry to the study of phytochrome is its quantification by immunoassay. Using the criterion of R/FR photoreversibility, phytochrome has been shown to control a wide range of developmental phenomena in plants. The simple concept of phytochrome action arising from such experiments is that only Pfr is physiologically active and the degree of response is related to the amount of Pfr. The logical experimental approach to test this hypothesis has been to attempt to correlate Pfr concentration with the degree of photoresponse by making parallel measurements on the phytochrome system along with the chosen test response. Studies on the biochemistry of phytochrome action also require that small quantities of phytochrome be measured routinely and accurately. Prior to the advent of immunoassays, the only method available for the detection and measurement of phytochrome was a direct spectrophotometric assay based on the spectral differences between Pr and Pfr (Pratt, 1979). Spectrophotometric assays have a lower detection limit in the region of 350 ng cm^{-3}, which is sufficient for the measurement of phytochrome in extracts of etiolated seedlings. However, when etiolated plants are transferred to light, the concentration of phytochrome drops rapidly to below the limit of spectrophotometric detection. Furthermore, spectrophotometric measurements are subject to gross interference by chlorophyll so that even if the level of phytochrome in light-grown plants was similar to that in the dark they would still not be possible in green plant tissue. In consequence, almost all studies on the phytochrome system have involved the use of etiolated seedlings, with the assumption that findings will be applicable to light-grown plants. Hunt & Pratt (1979) were the first to devise an immunoassay which overcame the limitations of spectrophotometric measurements (Fig. 1). Their radioimmunoassay (RIA) had a sensitivity limit of approximately 2 ng and was not subject to interference by chlorophyll. Using this assay, these authors were able to make phytochrome measurements in extracts from green, light-grown oat seedlings (Hunt & Pratt, 1980*b*) and follow changes in apparent phytochrome content in light-dark cycles. Although this should have represented a major breakthrough in phytochrome measurement, the assay itself was complicated, slow and difficult to carry out. It was not widely adopted and only one other example of phytochrome measurement by RIA has been published (Gottmann & Schafer, 1983). Recently, we published details of a double-antibody sandwich enzyme-

Fig. 1. Principles of the phytochrome RIA and sandwich ELISA. RIA is based on the displacement of radioactively labelled phytochrome from the antiphytochrome IgG by unlabelled phytochrome present in plant extracts. Experimental details are in Hunt & Pratt (1979). Sandwich ELISA is based on the immunoabsorption of phytochrome from a plant extract and its detection by a subsequent incubation with an excess of antiphytochrome immunoglobulin which may be enzyme-labelled directly (a) or indirectly by a second antibody (b). Practical details of the sandwich ELISA for phytochrome are given in the Protocols.

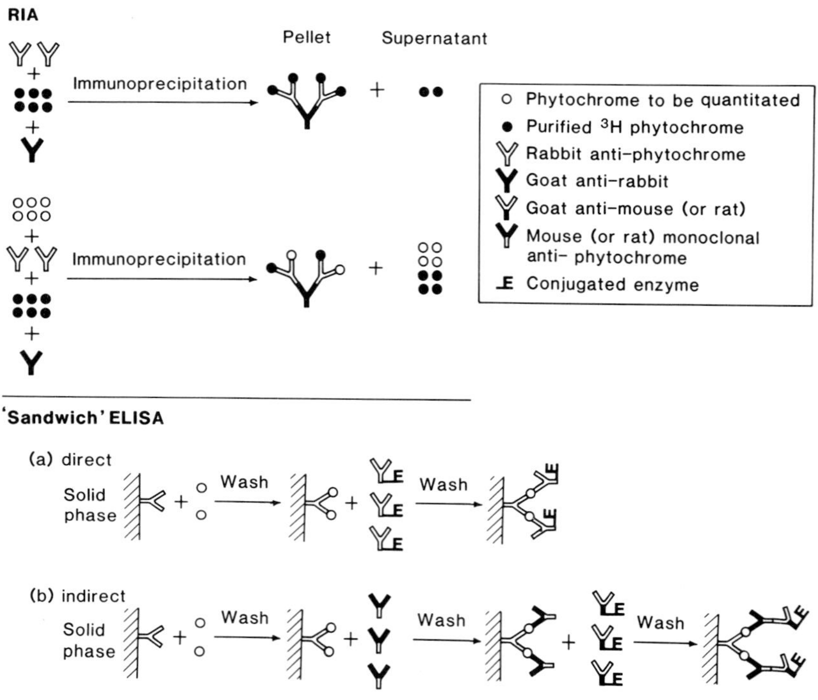

linked immunosorbent assay (ELISA) for phytochrome using rabbit antiserum raised against large phytochrome from etiolated oat seedlings (Thomas, Crook & Penn, 1984*a*). A similar assay was also developed independently in Pratt's laboratory using a combination of rabbit antisera and mouse monoclonal antibodies (McAb) to oat phytochrome (Shimazaki, Cordonnier & Pratt, 1983). These 'sandwich' assays (Fig. 1) are at least equivalent to RIA in terms of sensitivity and susceptibility to interference by chlorophyll. Unlike RIA, however, they give rapid, reproducible results and are extremely simple to perform. When we attempted to use our ELISA to measure phytochrome in extracts of light-grown oat seedlings, we observed that dilution curves had a reduced slope and saturated at a lower level of enzyme activity than those for dark extracts. Mixing experiments showed that these differences were not caused by inhibitors in the extract and we concluded that phytochromes from light- and dark-grown oat seedlings were antigenically different (Thomas *et al.*, 1984*a*). Shimazaki *et al.*, (1983) also noted discrepancies in their ELISA between the reactivity of phytochrome from etiolated seedlings and phytochrome from those which had been transferred to light. We have since confirmed these differences by an ELISA using rat McAb (Fig. 2) and have also observed similar results in wheat (*Triticum aestivum*). These observations are

Fig. 2. Concentration-response curves for indirect ELISA of phytochrome in extracts of oat seedlings grown in darkness or transferred to red light for 5 h prior to extraction. Phytochrome concentration was determined by spectrophotometry. ELISA measurements used monoclonal antibody AFRC MAC 56. Note that after 5 h red light the phytochrome present shows less activity (antigenicity) in ELISA.

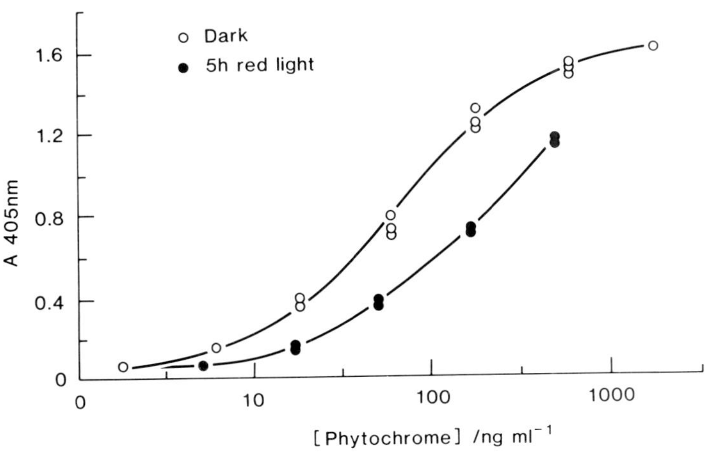

consistent with proposals made on the basis of spectrophotometric measurements and phytochrome mRNA determinations that two pools of phytochrome which differ in their rates of turnover exist in oat seedlings (Gottmann & Schafer, 1983; Quail *et al.*, 1983). Phytochrome which accumulates in etiolated seedlings (dark phytochrome) is rapidly degraded as Pfr on transfer to light leaving a residual pool of phytochrome that is apparently stable (light phytochrome). Using a different approach, Tokuhisa, Daniels & Quail (1984) partially purified phytochrome from light-grown oat seedlings and found differences in antigenicity, molecular size and peptide map between it and dark phytochrome. In both our own and Pratt's ELISAs the antibodies used were raised to phytochrome from etiolated seedlings which is predominantly dark phytochrome. Standard curves were constructed using purified dark phytochrome or extracts from etiolated seedlings. Antigenic differences between light and dark phytochrome would inevitably lead to discrepancies in the ELISA measurements when plants are transferred from dark to light as the relative amounts of the two types of phytochrome would be greatly altered. The uncertainty regarding the calibration of these assays makes it impossible to interpret the published RIA and ELISA data on changes in phytochrome content in a strict quantitative sense. Both light and dark phytochrome have similar spectral characteristics and are not distinguished in spectrophotometric measurements. The immediate challenge is therefore to develop quantitative immunoassays which are specific and calibrated for particular types of phytochrome.

One major disadvantage of immunoassays as compared to spectrophotometric assays is that only total phytochrome is measured and no distinction between Pr and Pfr is made. This is true for all assays which use antisera as these preparations do not show any differences in net binding between Pr and Pfr. Specific assays would, however, be possible if McAb which discriminated between Pr and Pfr could be found. We have prepared a set of McAb to large dark oat phytochrome which includes some which react differentially with Pr and Pfr (Thomas *et al.*, 1984*b*). These antibodies show relatively low affinity for phytochrome and for this reason are not ideal for immunoassays. Nevertheless the fact that discriminating antibodies exist indicates that this approach is feasible and merits further investigation.

Cross-reactivity of antibodies

Phytochrome is a large protein and although all phytochrome molecules probably have a common ancestry, variations in the amino acid composition of phytochrome from different species can be expected to have occurred during evolutionary divergence. One means of investigating homology between phytochrome molecules is to determine the extent to which phytochrome from a given species cross-reacts with antibodies raised to phytochrome from a different species. Comparative studies using antisera preparations raised against phytochrome from both monocotyledonous and dicotyledonous species have been carried out by Pratt and co-workers (Pratt, 1973; Cordonnier and Pratt, 1982). They found that antibodies raised to oat phytochrome reacted strongly in double diffusion assays with phytochrome from other members of the Gramineae, but less well with phytochrome from dicotyledonous species. Conversely, antibodies to phytochrome from lettuce (*Lactuca sativa*), zucchini (*Cucurbita pepo*) and pea (*Pisum sativum*) reacted more strongly with dicotyledonous phytochromes and only slightly or not at all with monocotyledonous phytochrome. When dicotyledonous species were compared, pea, bean (*Phaseolus vulgaris*) and lettuce phytochrome were antigenically similar, whereas zucchini phytochrome appeared to be less closely related.

The interest in phytochrome cross-reactivity is not limited to its usefulness in establishing the taxonomic relationship between species. The expectation that parts of the phytochrome molecule important for its function are likely to be highly conserved has prompted the search for antibodies which cross-react with phytochrome from a range of species. This line of investigation has been vigorously pursued by Cordonnier and Pratt (Cordonnier & Pratt, 1982; Cordonnier, Greppin & Pratt, 1984) who have identified antibodies with a broad range of cross-reactivity in both polyclonal and monoclonal preparations. Considerable care should, however, be taken in the interpretation of the results of such studies. The combining site for a particular antibody consists of only a few amino acids and it is possible to envisage that changing one of these could result in a total loss of antigenicity but have only a small effect on other properties of that particular region of the protein. Lack of cross-reactivity, therefore, does not necessarily mean that an antibody binds to a part of the phytochrome which is functionally unimportant. Conversely, conservation of short sequences of amino acids may be simply fortuitous. For these reasons cross-reactivity studies are likely to be most enlightening when carried out as part of a broader investigation of the properties of particular antibodies. A recent example of this is the identification by Daniels and Quail (1984) of a McAb raised to oat phytochrome which bound to phytochrome from zucchini. This antibody was shown to recognize an antigenic domain which lies within 6 kDa of the amino terminus and is thought to be important for correct protein-chromophore interaction in phytochrome from a wide range of species (Vierstra *et al.*, 1984).

Highly conserved epitopes also have another significance in that they may allow the development of methods of immunoassay and immunopurification that could be used in a wide range of species. This would represent a considerable advance, as currently a species for study is often selected on the basis of how easy it is to measure and characterize its phytochrome rather than the importance of its light responses.

Immunocytochemical localization of phytochrome

Studies based on the application of immunocytochemical techniques have provided important information on the distribution of phytochrome at both tissue (Pratt & Coleman, 1971, 1974) and sub-cellular levels (Verbelen *et al.*, 1982; Saunders *et al.*, 1983). The objective of the technique is to obtain thin sections from fixed and embedded plant material so that the integrity of the cellular organization is preserved. The particular antigen of interest (phytochrome) is then located by a specific antibody and subsequently revealed by conjugation of that antibody to a detection system (e.g. an enzyme reaction or fluorescent chemical for light microscopy or ferritin label for electron microscopy). For the successful application and interpretation of results from this technique it is essential to ensure that all the reagents used for the detection are able to penetrate the tissue completely. This criterion has not always been satisfied, although refinements in methodology have improved this aspect.

In the early experiments, the sectioned tissue was treated sequentially with rabbit anti-phytochrome serum, sheep anti-rabbit immunoglobulin serum and purified rabbit peroxidase-antiperoxidase complex. The indirectly linked peroxidase was then visualized with 3,3′-diaminobenzidine in buffered hydrogen peroxide and the oxidized substrate revealed under the light microscope as an orange-brown stain. By this method high concentrations of phytochrome were detected in a variety of non-irradiated tissue, including coleoptile tip, shoot apex, root cap and hydrated seeds (Pratt & Coleman, 1974; Coleman & Pratt, 1974*a*). At the sub-cellular level phytochrome was found, in most cases, to be uniformly distributed throughout the cytoplasm in etiolated tissue. These general findings were confirmed and extended by the use of electron microscopy (EM) (Coleman & Pratt, 1974*b*). One important observation arising from these early studies was that *in vivo* photoconversion of Pr to Pfr gave rise to a sub-cellular redistribution into discrete sequestered areas (Mackenzie *et al.*, 1975). No specific organelle could be identified in relation to this sequestered phytochrome population. After photoconversion back to Pr, phytochrome slowly resumed its uniform distribution (within 2 h). The immunocytochemical studies were also consistent with an increase in the proportion of phytochrome associated with membranes from 5 to 60% following red-irradiation *in vivo* (Pratt, 1978). Fig. 3 illustrates the uniform distribution of phytochrome in non-irradiated tissue and the subsequent sequestration after a red light treatment.

The original immunocytochemical methods suffered from several technical difficulties which influenced both the structural integrity of the tissue and the immunological resolution that was obtained (Verbelen *et al.*, 1982). Epel *et al.*, (1980) overcame these difficulties by using immunoaffinity purified rabbit IgG instead of antisera and introduced cryosectioning to give thin sections (0.5–1.0 μm in contrast to the previous 3–10 μm). The phytochrome was revealed in the sections by consecutive incubations of rabbit anti-phytochrome IgG and rhodamine-conjugated goat anti-rabbit. Using these techniques on a variety of monocotyledonous species it was confirmed that phytochrome was uniformly distributed throughout the cytoplasm in non-irradiated tissue and, upon conversion to Pfr, was detected in discrete sequestered areas. The level of detectable immunofluorescence in the nucleoplasm varied between species. General agreement with these results was found by the use of EM on cryosectioned material (Verbelen *et al.*, 1982). The greater sensitivity of EM, however, did provide evidence that on irradiation the sequestered phytochrome appeared to be associated

Fig. 3. Immunofluorescence labelling of phytochrome with rhodamine in 1 μm thick cryosections of oat coleoptile tips. (a) Fluorescence image of a section of an etiolated plant showing diffuse distribution of antigen in the cytoplasm, (b) Comparable image from an illuminated (etiolated + 5 mm red light) plant; discrete areas of high fluorescence are visible against a low background cytoplasmic fluorescence. Picture courtesy of Dr. David McCurdy.

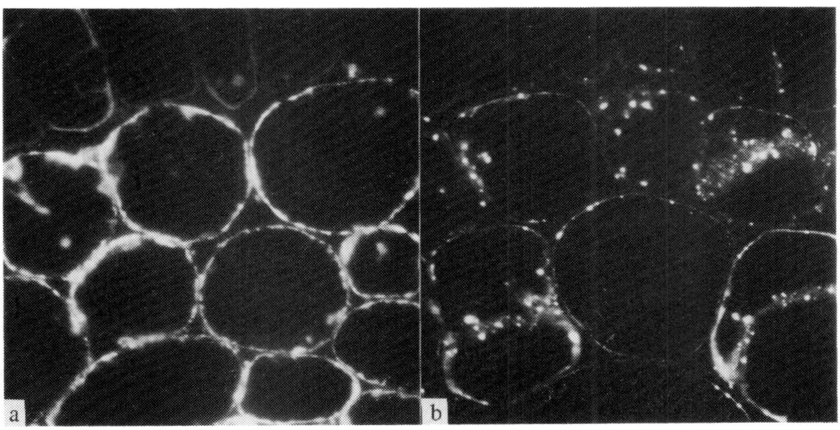

with small vacuoles and, to a lesser extent, in the large central vacuole. The suggestion has been put forward, therefore, that the redistribution of phytochrome occurs in order to incorporate it into the vacuolar system of the plant where it is subsequently broken down (MacKenzie *et al.*, 1975; Verbelen *et al.*, 1982). In addition, although phytochrome could not be detected in the nuclei of dark-grown oat, its presence was observed in plants exposed to the light, confirming the observations of Mackenzie, Briggs & Pratt (1978).

Immunocytochemical investigation of phytochrome distribution has been largely confined to monocotyledonous plants and procedures using antisera. Recently this approach has been extended to dicotyledonous plants by the use of McAb against pea phytochrome (Saunders *et al.*, 1983). Eight different McAb were used to visualize phytochrome in epicotyl sections and the reactivity of the monoclonals to purified pea phytochrome was also monitored by ELISA. Only 3 of the 8 McAb were found to be useful for the immunocytochemical localization, although all were strongly reactive to phytochrome when measured by the ELISA. Fixation of phytochrome on to the ELISA plate with 4% formaldehyde did indicate that this procedure could account for reduced binding to phytochrome in immunocytochemical studies. However, two McAb that bound well to formaldehyde-fixed phytochrome as assayed by ELISA showed a low immunocytochemical response. It is apparent from these results that extreme caution must be taken in the interpretation of immunocytochemical localization studies. Using the McAb that showed the strongest staining it was found, in agreement with the findings in monocotyledonous plants, that in non-irradiated tissue phytochrome was uniformly distributed throughout the cytoplasm. The cortical cells of the hook region of the epicotyl contained the highest level of phytochrome. Most epidermal cells did not show a marked amount of phytochrome but, in contrast, stomatal guard cells displayed an enhanced immunofluorescence (Fig. 4). This high level of phytochrome in guard cells may reflect its possible role in stomatal action (Holmes & Klein, 1983). In red-irradiated tissue phytochrome sequestering was observed only if the tissue was left in the dark for 10 min after irradiation and prior to fixation. Immediate fixation after red irradiation resulted in an inability to detect phytochrome. Both phenomena could be prevented or reversed by the appropriate far-red irradiation. The transient inability to detect phytochrome was thought not to be due to a conformational change but rather to an association with an as yet unidentified sub-cellular component that interfered with the antibody-antigen binding.

Membrane association of phytochrome

There is considerable indirect physiological evidence that suggests that the molecular mechanism of phytochrome action is mediated via an interaction with cell membranes (Marmé, 1977; Brownlee, Roth-Bejerano & Kendrick, 1979). However, these results contrast with the apparent water soluble properties of phytochrome and its largely cytosolic distribution in non-irradiated plant cells (Quail, 1981). The accurate assessment of phytochrome associations with sub-cellular fractions has undoubtedly been hindered by the use of the spectrophotometric assay. This method is limited in its sensitivity and is affected by the light-scattering properties of membrane extracts. It is therefore surprising that alternative immunological approaches have only found limited application to date. In an attempt to enrich and subsequently identify a phytochrome-specific membrane fraction, Vierstra, Tokuhisa & Quail (1983) used solid phase antibodies (Ito & Palade, 1978). Four-day-old dark-grown oat seedlings were labelled prior to membrane isolation with either [^{14}C]-choline or $^{35}SO_4^{2-}$ so that the membrane distribution could be monitored. Red-irradiation *in vivo* and extraction in the presence of 10 mM Mg^{2+} induced phytochrome to associate with a 50,000 g × 5 min fraction. Particles with associated phytochrome were immunoprecipitated from the resuspended membrane pellet following incubation with purified monospecific anti-phytochrome IgG and heat-killed *S. aureus* cells. By this procedure, although all the phytochrome was found to be associated with the solid phase antibodies, only 5% of the label originally present in the isolated membranes was

Fig. 4. Paired phase contrast (a) and fluorescence (b) micrographs of dark-grown pea epicotyl epidermal cell labelled with monoclonal antibody 11–18 a 4. g = guard cell. Picture from Saunders *et al.*, 1984.

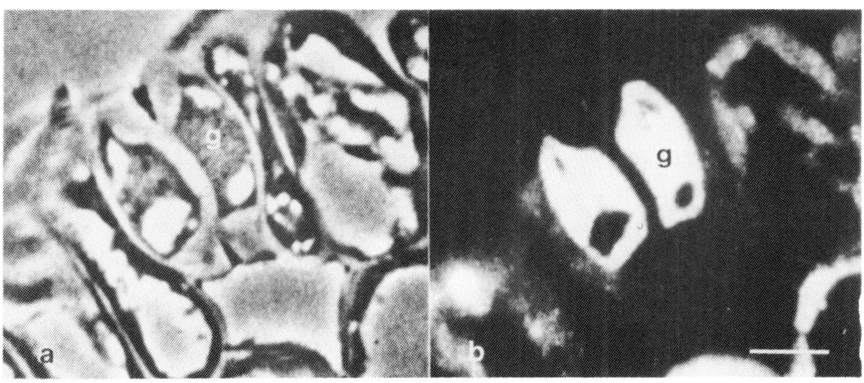

immunospecifically absorbed. This 20-fold enriched membrane fraction was then analysed by EM, enzyme markers and 2-dimensional electrophoresis. The results indicated that no unique phytochrome-bearing membrane fraction could be identified but that phytochrome is associated with all membrane fractions.

Jordan, Partis & Thomas (1984) used ELISA and Western blotting to study the phytochrome-membrane association. Membranes were isolated by differential centrifugation and ELISA clearly demonstrated that phytochrome was associated with membranes isolated from non-irradiated tissue in addition to the enhanced association found in red-light-irradiated tissue. Immunological differences were detected between membranes and the post-microsomal soluble fraction from etiolated tissue, similar to those found between phytochrome present in green and dark-grown tissue (Thomas *et al.*, 1984*a*). This suggested that the small percentage of phytochrome associated with these membranes (ca. 5%–10% Rubinstein *et al.*, 1969; Jordan, 1982) was not simply the result of co-precipitation of soluble phytochrome with membranes during fractionation. Western blotting of the membrane-associated phytochrome following denaturing sodium dodecylsulphate polyacrylamide gel electrophoresis (SDS-PAGE) demonstrated that its molecular size was equivalent to that of soluble phytochrome (124 kDa). Some evidence for the specific association of phytochrome with oat mitochondria has been presented by Roux *et al.*, (1982).

Molecular biology of phytochrome

The application of immunology in conjunction with molecular biological techniques has in recent years increased the understanding of both the molecular properties of phytochrome and the regulation of its synthesis. Evidence that the monomer molecular weight of phytochrome was 124 kDa and not 118 kDa as previously thought, was obtained by *in vitro* translation studies (Bolton & Quail, 1982; Gottman & Schafer, 1982). A single polypeptide was immunoprecipitated by monospecific antiphytochrome IgG from the translation products of either rabbit reticulocyte or wheat-germ protein synthesizing systems programmed by oat polyadenylated RNA. The immunoprecipitate obtained was then subjected to SDS-PAGE and had a molecular weight of 124 kDa, which has been independently confirmed by other procedures (Vierstra & Quail, 1982*a*).

A number of workers have recently investigated the levels of mRNA coding for phytochrome using *in vitro* translation systems and subsequent immunoprecipitation of the translation products with antisera (see Cuming *et al.*, this volume, for a description of the method) to phytochrome (Gottman & Schafer, 1982; Gottman & Schafer 1983; Otto *et al.*, 1983;

Colbert, Hershey & Quail, 1983). From these experiments it has been clearly demonstrated that phytochrome mRNA is a low abundance message in etiolated oat seedlings (0.005% of total translation products) (Colbert *et al.*,

Fig. 5. Identification of a phytochrome cDNA clone (pAP-3) by hybridization-selection and translation. Total poly (A)+ RNA or mRNA selected by hybridization to cDNA clone pAP-3 (containing phytochrome sequences) or pL 99 (a plasmid containing a cDNA insert similar in size to pAP-3) was translated in a rabbit reticulocyte cell-free protein-synthesizing system. Aliquots of each reaction mixture were withdrawn for analysis and the remainder immunoprecipitated with anti-phytochrome or non-immune serum. Translation products were separated by SDS-PAGE and fluorographed. Lanes A–C, total translation products; lanes D, E and G antiphytochrome immunoprecipitates; lane F, non-immune control immunoprecipitate; lanes A and D total poly(A)+RNA; lanes B, E, F mRNA selected by pAP-3; lanes C and G mRNA selected by P199 (from Hershey *et al.*, 1984).

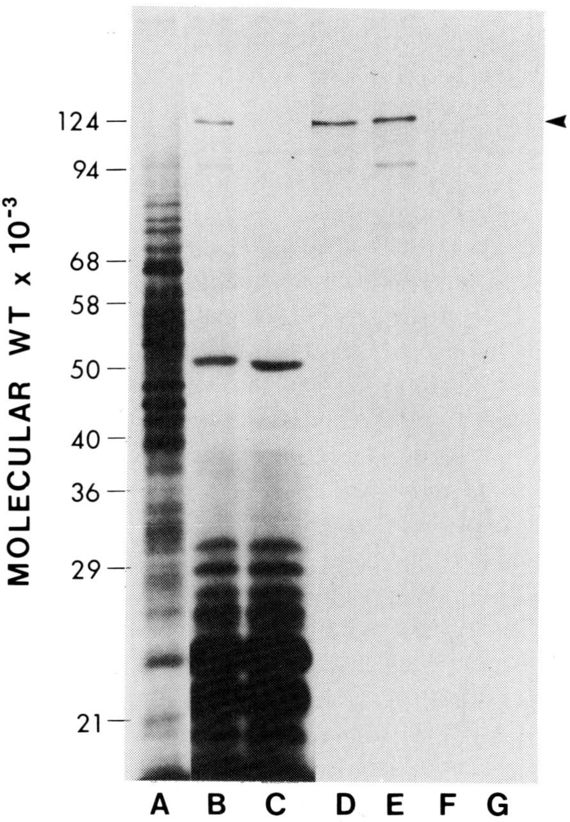

1983; Quail *et al.*, 1983), and is further reduced by light. A red-light pulse lowered the level of mRNA in etiolated tissue by 95% within 2 h and this effect was fully reversed by an immediate far-red treatment. Converting 1% of Pr to Pfr resulted in a 40% reduction in mRNA and saturation of the response required only 20% photoconversion to Pfr. The decline in phytochrome mRNA has also been shown to be a consequence of loss of phytochrome mRNA sequences rather than some form of translational control (Hershey *et al.*, 1984). It is apparent therefore that phytochrome can regulate the transcription of its own gene in etiolated oat seedlings (Quail, 1984).

Recently, the preparation of copy (c) DNA clones for phytochrome have been described (Hershey *et al.*, 1984) and immuno-probes have been used to demonstrate the specificity of the cDNA sequences. To screen for plasmids that contained cDNA for phytochrome mRNA, hybridization-selection followed by translation was used. Polyadenylated RNA from etiolated oat was hybridized to filter-bound plasmid containing presumptive cDNA for phytochrome. Bound mRNA was eluted from the filters and translated in a rabbit reticulocyte cell-free protein synthesizing system. Anti-phytochrome serum was then used to precipitate phytochrome. (See Cuming *et al.*, this volume, for further details of the methods.) Fig. 5 illustrates the identification of a cDNA clone for phytochrome using hybridization-selection and translation.

Antibodies as molecular probes

Although it is generally assumed that phytochrome exists as a dimer of two 124 kDa proteins, several lines of evidence suggest that this may be an oversimplification. In general, estimates of molecular weight obtained by electrophoresis have been at variance with those obtained by gel filtration or density centrifugation and this has been ascribed to the influence of hydrodynamic factors. If extracts of phytochrome are not heated in SDS prior to SDS-PAGE, bands corresponding to dimers, trimers, tetramers and traces of higher oligomers can be detected by Western blotting and visualization with either antisera or McAb (Partis, Jordan & Thomas, 1984). It has been shown that mild treatments with detergents (including SDS) do not affect phytochrome structure (Kim, Kim & Song, 1983). ELISAs in which plates were coated with a McAb to phytochrome and developed using alkaline phosphatase conjugated to the same monoclonal gave positive results, indicating that at least two copies of the epitopes are present (Partis *et al.*, 1984). While the possibility that there are repeating epitopes on the phytochrome monomer cannot be ruled out, the evidence suggests that phytochrome exists as at least a dimer, with a strong possibility of higher oligomers being the naturally occurring form.

With the advent of McAb to phytochrome from monocotyledonous and dicotyledonous plants (Cordonnier *et al.*, 1983; Nagatani *et al.*, 1983; Thomas *et al.*, 1984*b*; Daniels & Quail, 1984), the possibility of identifying and characterizing localized regions of the phytochrome molecule was realized. One approach has been to combine the use of McAb with the action of endogenous proteases to locate the domains on which particular binding sites or epitopes are located. Two strategies have been used in the characterization of the binding sites of the monoclonals. Purified preparations of small and large phytochrome have been used as antigens in ELISA to screen collections of McAb to oat, rye and pea phytochrome (Nagatani *et al.*, 1983; Cordonnier *et al.*, 1984). All of the antibodies screened in this way bound to small phytochrome, indicating that the epitopes for these monoclonals were located on the chromophore-containing domain. Although this procedure allows the screening of all monoclonals, it requires the prior purification of each of the peptide fragments. A second method has utilized the powerful techniques of SDS-PAGE and Western blotting followed by detection of immunoreactive peptides with alkaline phosphatase- or peroxidase-linked McAb. As the kinetics of proteolysis are well-documented (Quail *et al.*, 1983), highly purified preparations are not required. Quail (1984) has identified three regions of the phytochrome molecule which contain epitopes, a 6–10 kDa region proximal to the N-terminus, the 60 kDa chromophore-containing domain and a region of the 55 kDa C-terminal peptide. He has classified the monoclonals binding to these areas as type 1, 2 and 3 respectively (Daniels & Quail, 1984). Under this classification, the monoclonals produced by Nagatani *et al.*, (1983) and by Cordonnier *et al.*, (1983) are all type 2, and the McAb characterized so far in our laboratory are of types 2 and 3. A major disadvantage of this procedure is that only those McAb which recognize contiguous sequences on the primary sequence can be classified; for example a large number of hybridoma culture supernatants from our fusions contain monoclonals which do not recognize SDS-denatured phytochrome, despite giving high signals in assays such as ELISA, where the protein is retained in its native conformation.

Further resolution of the epitopes on phytochrome requires that alternative digestion procedures must be employed. We have used tryptic and chymotryptic digestion of SDS-denatured protein (Cleveland *et al.*, 1977) and digestion of urea/acetic acid denatured phytochrome with N-chlorosuccinimide (Lischwe & Ochs, 1982) to promote cleavage at arginine and lysine, phenylalanine and tyrosine, and tryptophan respectively. Peptides produced by these cleavage agents were separated by SDS-PAGE and transferred to nitrocellulose by Western blotting (see Protocols for further details). Immunoreactive peptides detected by the monoclonal antibodies enabled ordering of the epitopes along the primary sequence of phytochrome (Partis & Thomas, 1985). An added advantage of this approach is that large numbers of monoclonals may be screened for identity of binding site by simply comparing the patterns of peptide recognition.

The active form of phytochrome is Pfr and, presumably, the mode of action of phytochrome depends on the ability of the cell to recognize some essential difference between the two spectral forms. A change in the conformation of phytochrome during phototransformation is an attractive proposition and the known differences in physico-chemical properties (Smith, 1983) lends support to this hypothesis. Some of the evidence must be treated with caution, however, as experiments carried out in the absence of information on the size of the protein used may reflect only the differential susceptibility of Pr or Pfr to the action of protease. This situation is well exemplified by the report that Pr and Pfr behaved differently in microcomplement fixation (Hopkins & Butler, 1970). However, Pratt (1973), repeating the microcomplement fixation experiments using antibody to small phytochrome found that higher concentrations of large phytochrome were required for maximum activity compared to small phytochrome and Cundiff & Pratt (1975*a*), using antisera to large phytochrome, found that large phytochrome had a higher activity than small phytochrome. It is probable, therefore that the behaviour of Pr and Pfr in microcomplement fixation was due to differential proteolysis of Pr, rather than to any immunological difference between them. The best documented difference between Pr and Pfr is that of susceptibility to the action of protease. Cundiff & Pratt (1975*b*) using antisera to both small and large phytochrome found that different products were formed when phytochrome was incubated with endogenous protease as Pr or as Pfr. Four bands were found in immunoelectrophoresis of the fragments of Pr, but only three bands were found for Pfr. The studies of Vierstra & Quail (Vierstra & Quail 1982*a*, 1982*b*; Quail *et al.*, 1983) have led to the production of a cleavage map which describes proteolytic sites which are more vulnerable to attack as Pr or Pfr. It is possible that conformational differences between Pr and Pfr arise from alterations in protease-labile interdomain regions. When phytochrome is digested as Pr then two sites at the N-terminus are accessible to attack, releasing 6–10 kDa of the N-terminus. There is evidence that loss of this segment of the protein results in the chromophore becoming more accessible to the aqueous environment and it has been suggested that the function of the N-terminal domain is to protect the chromophore (Hahn *et al.*, 1984; Quail, 1984). Whether or not this is the case, there must be a conformational change at or around this site which allows access of protease or exposes the labile sequence. There is also an area in the middle of the primary sequence which is more labile in Pfr (Quail, 1984). The most attractive prospect for probing the differences between Pr and Pfr is the identification of McAb which can discriminate between the two spectral forms. Such McAb are not abundant; a screen of culture supernatants from 40 hybridoma cell lines by peptide mapping and by ELISA has identified only three isolates which show strong preferential interaction with Pfr and only one which shows positive discrimination towards Pr (Thomas *et al.*, 1984*b*; Partis & Thomas, 1985). To date, these represent the only discriminating antibodies which have been isolated. The ordering of epitopes by peptide mapping and Western blotting of phytochrome fragments produced by endogenous proteases have shown that the portion of the primary sequence along which the epitopes for these discriminatory peptides lie is that which is preferentially sensitive to proteolysis in Pfr (Fig. 6). The McAb which bind strongly to Pfr are located on the N-terminal (chromophore-bearing) side of this cleavage point, while the remaining McAb lie on the other.

Fig. 6. Map of the proposed location of binding sites for monoclonal anti-phytochrome antibodies on dark phytochrome from *Avena sativa*. AFRC MAC 50 binds more strongly to Pfr and MAC 49 binds more strongly to Pr. Also shown are the cleavage points for endogenous proteases and the form (Pr or Pfr) in which they are more readily cleaved.

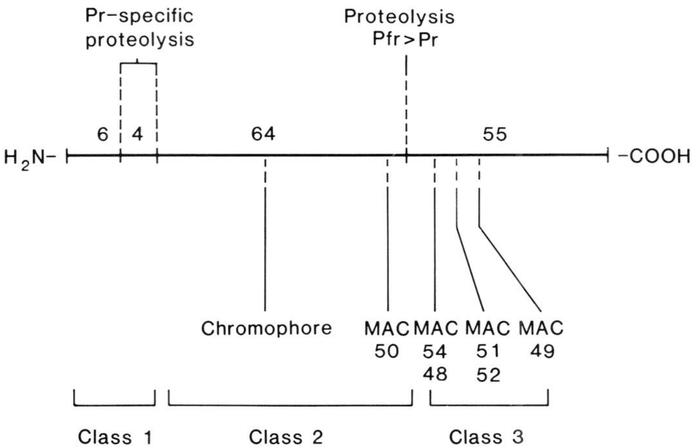

PROTOCOLS

I. Western blotting of protein digests

One of the most useful techniques we have used involves determining the pattern of peptide recognition when different monoclonal antibodies are reacted with partially hydrolysed proteins. Examination of such patterns makes it possible to identify antibodies that bind to closely adjacent sites, making it a useful screening procedure. Furthermore, analysis of combinations of antibodies which bind to particular fragments allows the ordering of epitopes along the protein and hence the construction of epitope maps (Partis & Thomas, 1985). In this section we describe methods for partial hydrolysis of phytochrome by three methods with different specificities (the latter allowing the separation of large protease-stable domains) and staining of nitrocellulose for peptide recognition. The buffers used for immunostaining of protein on nitrocellulose contain sodium azide as a preservative: take appropriate safety measures. The conditions for digestion and blotting are described for phytochrome but should require minor modification to apply to other protein antigens (see Cuming *et al.*, this volume, for a further procedure). Protein concentrations between 200 μg cm^{-3} and 1 mg cm^{-3} should prove satisfactory. The optimum concentrations of cleavage agent should be determined by prior experiment.

SDS-PAGE of the digested protein is essentially as described by Laemmli (1977) in 12.5% polyacrylamide gels but sodium mercaptoacetate is added in the electrophoresis buffer at 0.01% to prevent re-oxidation of reduced proteins. Polyacrylamide gradient gels may provide better resolution of digests than single concentration gels. Electrophoretic transfer of antigens from SDS gels on to nitrocellulose is as described by Burnette (1981) at 30 V cm^{-1} for 16 hours (see also Cuming *et al.*, this volume).

Digestion of denatured phytochrome with N-chlorosuccinimide (NCS)

1. Dissolve the protein in urea/acetic acid/water (1:1:1, w/v/v).
2. Incubate with NCS at 25 °C for 20 minutes as described by Lischwe and Ochs (1982)[a].
3. Terminate digestion by precipitation of peptides with 5 volumes acetone.
4. Wash the precipitated protein with 80% (v/v) acetone and dissolve in SDS sample buffer prior to electrophoresis.

Proteolytic digestion of denatured phytochrome

1. Boil the protein for two minutes in 50 mM Tris-HCl, 0.5% (w/v) SDS, pH 6.8.
2. Incubate the solubilized protein at 35 °C in the presence of the appropriate protease for 30 minutes[b].

3. Stop the digestion by the addition of 2-mercaptoethanol to 1% (v/v), SDS to 2% (w/v) and heating to 100 °C for 2 minutes.

Digestion of undenatured phytochrome by endogenous protease

1. Homogenize 5-day old etiolated *Avena* seedlings in 250 mM sucrose, 25 mM MOPS, 3 mM EDTA, 14 mM 2-mercaptoethanol pH 7.5.
2. Centrifuge the homogenate at 100,000 g for 60 minutes.
3. Illuminate the clarified extract with either red (660 nm) light or far-red (720 nm) light for 2 minutes.
4. Incubate at 25 °C for 6 hours[c].
5. Terminate the digestion of the extract by endogenous proteases by boiling for 2 minutes in 1% SDS.

Staining of nitrocellulose-bound antigen

1. Incubate the nitrocellulose sheet with 2% BSA in tris buffered saline (TBS) for 30 minutes at room temperature[d,e].
2. Dilute the primary antibody to a working concentration of 1–2 μg cm^{-3} with 1% BSA in PBST[f].
3. Incubate the diluted antibody with the sheet for 30 minutes at ambient temperature.
4. Remove the sheet and store the antibody at 0 °C with 0.02% sodium azide.
5. Remove non-specifically bound antibody from the sheet by three washes with PBST for five minutes each.
6. Incubate the washed sheet with species-specific peroxidase-conjugated anti-IgG in PBST containing 1% BSA for 30 minutes at room temperature to detect the antibody–antigen complex[g].
7. Remove the sheet and store the conjugate at 0 °C.
8. Wash the sheet three times in PBST for 5 minutes at room temperature to substantially reduce the level of background staining due to non-specific absorption of the second antibody.
9. Remove excess antibody and Tween by washing the sheet in TBS for five minutes[h].
10. Immerse the sheet in a solution of 4-chloro-1-naphthol in TBS/methanol plus 0.6 cm^3 hydrogen peroxide[i].
11. Wash the sheet thoroughly with distilled water and blot dry with filter paper once the bands reach the desired intensity[j].

II. Measurement of phytochrome by ELISA

The following is the method we use for measuring phytochrome by ELISA. A method for direct ELISA using antisera to immunoadsorb, and labelled antisera to immunolabel is presented. A variation of this method using a monoclonal antibody to label the phytochrome indirectly is also described. The assay is carried out in 96-well polystyrene plates (e.g. Nunc Immunoplate 1, GIBCO Ltd). Enzyme-labelled polyclonal anti-body was prepared as described in Thomas et al., (1984*a*). All volumes refer to quantities added to each well. All buffers contain 0.02% sodium azide as preservative. Take appropriate safety precautions.

Two methods can be used, EITHER the direct (steps 1–16) OR indirect (steps 17–24) techniques.

EITHER

1. Dilute antibody (anti-phytochrome antiserum IgG) with sufficient coating buffer (sodium carbonate, pH 9.6) to fill each well of the plate with 100 μl of solution at 4 μg cm^{-3} antibody.
2. Incubate plate at room temperature for 2 hours.
3. Remove coating buffer and wash plate with PBST three times for five minutes.
4. Incubate plate with 150 μl bovine serum albumin in PBST for 1 hour at room temperature to block any remaining protein-binding sites.
5. Wash plate twice with PBST for 5 minutes, blotting plate on paper towels.
6. Add 100 μl phytochrome-containing solution diluted to appropriate concentration in PBST containing 1% BSA.
7. Incubate for 2 hours at 0–4 °C.
8. Remove sample solutions and blot on paper towels.
9. Wash three times with PBST for five minutes, blotting plate on towels.
10. Add 100 μl of enzyme-antibody conjugate diluted 1:500 with PBST containing 1% (w/v) BSA.
11. Incubate for 2 hours at room temperature.
12. Remove conjugate, wash plate three times with PBST for five minutes and blot.
13. Add 100 μl p-nitrophenyl phosphate (1 mg cm^{-3} in 10% diethanolamine pH 9.8).
14. Incubate at room temperature until colour develops.
15. Stop colour development with 25 μl 3M sodium hydroxide (optional step).
16. Read colour at 405 nm.

OR

17. Steps 1–7 as direct method.
18. Add 100 μl of suitably diluted rat (or mouse) McAb in PBST containing 1% BSA.
19. Incubate for 1 hour at room temperature.
20. Remove McAb and blot plate on a paper towel.
21. Wash plate three times for five minutes in PBST.
22. Add 100 μl of phosphatase-labelled anti-rat antibody (Miles, U.K.) diluted 1:1500 in PBST containing 1% BSA.
23. Incubate for 1 hour at room temperature.
24. Steps 12–16 of the direct method.

Notes

a For digestion of phytochrome, we use a concentration of 0.5 mg cm^{-3} NCS.
b For phytochrome, we use a concentration of 10 or 100 μg cm^{-3} respectively of either chymotrypsin or trypsin.
c To increase the concentration of antigen, purified phytochrome may be added prior to irradiation.
d This saturates the protein binding capacity of the nitrocellulose before the addition of the antibody.
e If the Western blot is to be stained with Amido Black to visualise all the proteins, the blot should be incubated with 0.2% Tween 20 in phosphate buffered saline (PBST) instead of BSA. Additionally, the sheet may be incubated with serum from the same source as the peroxidase-conjugated anti-IgG (see step 6) to reduce interference by serum-binding proteins.
f Typically, the primary antibody should be at ca. 1–2 μg cm^{-3}, although this will depend on the affinity of the antibody for the antigen and on the purity of the antibody or antiserum.
g The working dilution of the second antibody is typically 1:1000 but lower dilutions may be required. Omit azide from this and subsequent steps.
h This step is essential to prevent the precipitation of the substrate in step 10.
i The substrate is not very soluble in aqueous buffer and will precipitate if the concentration of methanol is too low or if detergent e.g. Tween is present. To make the chloronaphthol solution, weigh out 30 mg and dissolve it in 10 cm^3 of methanol. When the chloronaphthol is completely dissolved, add 30 cm^3 TBS. The stain should develop in 5–15 minutes.
j Do not allow the sheet to wrinkle. The nitrocellulose will darken in direct light, so run staining in dim light and keep stained sheets in the dark.
k An example of the results obtained is shown in Fig. 2.

KEY REFERENCES

Colbert, J. T., Hershey, H. P. & Quail, P. H. (1983). Autoregulatory control of translatable phytochrome mRNA levels. *Proceedings of the National Academy of Sciences, U.S.A.*, **80**, 2248–52.

Cordonnier, M. M., Smith, C., Greppin, H. & Pratt, L. H. (1983). Production and purification of monoclonal antibodies to Pisum and Avena phytochrome. *Planta*, **158**, 369–76.

Hershey, H. P., Colbert, J. T., Lissemore, J. L. & Barker, R. F. (1984). Molecular cloning of cDNA for Avena phytochrome. *Proceedings of the National Academy of Sciences, U.S.A.*, **81**, 2332–6.

Kerscher, L. & Nowitzki, S. (1982). Western blot analysis of a lytic process in vitro specific for the red light absorbing form of phytochrome. *FEBS Letters*, **146**, 173–6.

Mackenzie, J. M., Coleman, R. A., Briggs, W. R. & Pratt, L. H. (1975). Reversible redistribution of phytochrome within the cell upon conversion to its physiologically active form. *Proceedings of the National Academy of Sciences, U.S.A.*, 72, **799–803**.

Viestra, R. D. & Quail, P. H. (1982*a*). Native phytochrome: inhibition of proteolysis yields a homogenous monomer of 124 kilodaltons from Avena. *Proceedings of the National Academy of Sciences, U.S.A.*, **79**, 5272–6.

REFERENCES

Boeshore, M. L. & Pratt, L. H. (1981). Characterisation of a molecular modification of phytochrome that is associated with its conversion to the far-red-absorbing form. *Plant Physiology*, **68**, 789–97.

Bolton, G. W. & Quail, P. H. (1982). Cell-free synthesis of phytochrome apoprotein. *Planta*, **155**, 212–17.

Brownlee, C., Roth-Bejerano, N. & Kendrick, R. E. (1979). The molecular mode of phytochrome action. *Science Progress, Oxford*, **66**, 217–29.

Burnette, W. N. (1981). "Western blotting": electrophoretic transfer of proteins from sodium dodecyl sulfate-polyacrylamide gels to unmodified nitrocellulose and detection with antibody and radioiodinated protein A. *Analytical Biochemistry*, **112**, 195–203.

Cleveland, D. W., Fischer, S. G., Kirschner, M. W. & Laemmli, U. K. (1977). Peptide mapping by limited proteolysis in sodium dodecyl sulfate and analysis by gel electrophoresis. *Journal of Biological Chemistry*, **252**, 1102–6.

Colbert, J. T., Hershey, H. P. & Quail, P. H. (1983). Autoregulatory control of translatable phytochrome mRNA levels. *Proceedings of the National Academy of Sciences, U.S.A.*, **80**, 2248–52.

Coleman, R. A. & Pratt, L. H. (1974*a*). Phytochrome: immunocytochemical assay of synthesis and destruction. *Planta*, **119**, 221–32.

Coleman, R. A. & Pratt, L. H. (1974*b*). Electron microscopic localisation of phytochrome in plants using an indirect antibody-labelling method. *Journal of Histochemistry and Cytochemistry*, **22**, 1039–47.

Cordonnier, M-M. & Pratt, L. H. (1982). Comparative phytochrome immunochemistry as assayed by antisera against both monocotyledonous and dicotyledonous phytochrome. *Plant Physiology*, **70**, 912–16.

Cordonnier, M. M., Smith, C., Greppin, H. & Pratt, L. H. (1983). Production and purification of monoclonal antibodies to Pisum and Avena phytochrome. *Planta*, **158**, 369–76.

Cordonnier, M. M., Greppin, H. & Pratt, L. H. (1984). Characterisation by enzyme linked immunosorbent assay of monoclonal antibodies to Pisum and Avena phytochrome. *Plant Physiology*, **74**, 123–7.

Correll, D. L., Steers, E., Towe, K. H. & Shropshire, W. (1968). Phytochrome in etiolated annual rye. IV. Physical and chemical characterisation of phytochrome. *Biochimica et Biophysica Acta*, **168**, 46–57.

Cundiff, S. C. & Pratt, L. H. (1973). Immunological determination of the relationship between large and small sizes of phytochrome. *Plant Physiology*, **51**, 210–13.

Cundiff, S. C. & Pratt, L. H. (1975*a*). Immunological and physical characterisation of the products of phytochrome proteolysis. *Plant Physiology*, **55**, 212–17.

Cundiff, S. C. & Pratt, L. H. (1975*b*). Phytochrome characterisation by rabbit antiserum against high molecular weight phytochrome. *Plant Physiology*, **55**, 207–11.

Daniels, S. M. & Quail, P. H. (1984). Monoclonal antibodies to three separate domains on 124 kilodalton phytochrome from Avena. *Plant Physiology*, **76**, 622–6.

Epel, B. L., Butler, W. L., Pratt, L. H. & Tokuyasu, K. T. (1980). Immunofluorescence localisation studies of the Pr and Pfr forms of phytochrome in the coleoptile tips of oats, corn and wheat. In *Photoreceptors and Plant Development*, ed. J. De Greef, pp. 121–133. Antwerpen: Antwerpen University Press.

Furuya, M. (1983). Molecular properties of phytochrome. *Philosophical Transactions of the Royal Society of London*, **303**, 361–75.

Gardner, G., Pike, C. S., Rice, H. V. & Briggs, W. R. (1971). "Disaggregation" of phytochrome in vitro – a consequence of proteolysis. *Plant Physiology*, **48**, 686–93.

Gottmann, K. & Schafer, E. (1982). In vitro synthesis of phytochrome apoprotein directed by mRNA from light and dark grown Avena seedlings. *Photochemistry and Photobiology*, **35**, 521–25.

Gottmann, K. & Schafer, E. (1983). Analysis of phytochrome kinetics in light-grown *Avena sativa* L. seedlings. *Planta*, **157**, 392–400.

Hahn, T. R., Song, P-S., Quail, P. H. & Vierstra, R. D. (1984). Tetranitromethane oxidation of phytochrome as a function of spectral form and molecular weight. *Plant Physiology*, **74**, 755–8.

Hershey, H. P., Colbert, J. T., Lissemore, J. L. & Barker, R. F. (1984). Molecular cloning of cDNA for Avena phytochrome. *Proceedings of the National Academy of Sciences, U.S.A.*, **81**, 2332–6.

Holmes, M. G. & Klein, W. H. (1983). Photocontrol of stomatal movement. In *Photomorphogenesis of Plants*, ed. C. Sundqvist, p. 22. Lund: University of Lund.

Hopkins, D. W. & Butler, W. L. (1970). Immunochemical and spectroscopic evidence for protein conformational changes in phytochrome transformations. *Plant Physiology*, **45**, 567–70.

Hunt, R. E. & Pratt, L. H. (1979). Phytochrome radioimmunoassay. *Plant Physiology*, **64**, 327–31.

Hunt, R. E. & Pratt, L. H. (1980*a*). Partial characterization of undegraded oat phytochrome. *Biochemistry*, **19**, 390–4.

Hunt, R. E. & Pratt, L. H. (1980*b*). Radioimmunoassay of phytochrome in green, light grown oats. *Plant Cell Environment*, **3**, 91–5.

Ito, A. & Palade, G. E. (1978). Presence of NADPH-cytochrome P-450 reductase in rat liver golgi membranes. *Journal of Cell Biology*, **79**, 590–7.

Jordan, B. R. (1982). Evidence for the interaction of membrane lipids and the plant photoreceptor, phytochrome. In *Biochemistry and Metabolism of Plant Lipids*, eds. J. G. F. M. Wintermans & P. J. C. Kuiper, pp. 479–82. Amsterdam: Elsevier Biomedical Press B.V.

Jordan, B. R., Partis, M. D. & Thomas, B. (1984). A study of phytochrome-membrane association using an enzyme-linked immunosorbent assay and Western blotting. *Physiologia Plantarum*, **60**, 416–21.

Kerscher, L. (1983). Subunit size, absorption spectra and dark reversion kinetics of rye phytochrome purified in the far-red-absorbing form. *Plant Science Letters*, **32**, 133–8.

Kerscher, L. & Nowitzki, S. (1982). Western blot analysis of a lytic process in vitro specific for the red light absorbing form of phytochrome. *FEBS Letters*, **146**, 173–6.

Kim, I-S., Kim, E. S. & Song, P-S. (1983). Interactions of phytochrome with detergents and liposomes. *Biochimica et Biophysica Acta*, **747**, 55–64.

Laemmli, U. K. (1977). Cleavage of structural proteins during the assembly of head of bacteriophage T4. *Nature*, **227**, 680–5.

Lischwe, M. A. & Ochs, D. (1982). A new method for partial peptide mapping using N-chlorosuccinimide/urea and peptide silver staining in sodium dodecyl sulfate-polyacrylamide gels. *Analytical Biochemistry*, **127**, 453–7.

Mackenzie, J. M., Coleman, R. A., Briggs, W. R. & Pratt, L. H. (1975). Reversible redistribution of phytochrome within the cell upon conversion to its physiologically active form. *Proceedings of the National Academy of Sciences, U.S.A.*, **72**, 799–803.

Mackenzie, J. M., Briggs, W. R. & Pratt, L. H. (1978). Intracellular phytochrome distribution as a function of its molecular form and of its destruction. *American Journal of Botany*, **65**, 671–6.

Marmé, D. (1977). Phytochrome: membranes as possible sites of primary action. *Annual Review of Plant Physiology*, **28**, 173–98.

Mumford, F. E. & Jenner, E. L. (1966). Purification and characterization of phytochrome from oat seedlings. *Biochemistry*, **5**, 3657–62.

Nagatani, A., Yamamoto, K. T., Furuya, M., Fukumoto, T. & Yamashita, A. (1983). Production and characterisation of monoclonal antibodies to rye (*Secale cereale*) phytochrome. *Plant Cell Physiology*, **24**, 1143–9.

Otto, V., Mosinger, E., Sauter, M. & Schaefer, E. (1983). Phytochrome control of its own synthesis in *Sorghum vulgare* and *Avena sativa*. *Photochemistry and Photobiology*, **38**, 693–700.

Partis, M. D., Jordan, B. R. & Thomas, B. (1984). Comparison of soluble and membrane phytochrome in Avena. *Plant Physiology Supplement*, p. 75.

Partis, M. D. & Thomas, B. (1985). Conformational changes in phytochrome determined by using monoclonal antibodies. *Biochemical Society Transactions*, in press.

Pratt, L. H. (1973). Comparative Immunochemistry of Phytochrome. *Plant Physiology*, **51**, 203–9.

Pratt, L. H. (1978). Molecular properties of phytochrome. *Photochemistry and Photobiology*, **27**, 81–105.

Pratt, L. H. (1979). Phytochrome: function and properties. In *Photochemical and Photobiological Reviews*, vol. 4, ed. K.C. Smith, pp. 59–124. New York: Plenum Press.

Pratt, L. H. & Coleman, R. A. (1971). Immunocytochemical localisation of phytochrome. *Proceedings of the National Academy of Sciences, U.S.A.*, **68**, 2431–5.

Pratt, L. H. & Coleman, R. A. (1974). Phytochrome distribution in etiolated grass seedlings as assayed by an indirect antibody-labelling method. *American Journal of Botany*, **61**, 195–202.

Quail, P. H. (1981). Intracellular location of phytochrome. In *Trends in Photobiology Laustriat*, pp. 485–500. New York: Plenum Press.

Quail, P. H. (1984). Phytochrome: a regulatory photoreceptor that controls the expression of its own gene. *Trends in Biochemical Science*, 450–3.

Quail, P. H., Colbert, J. T., Hershey, H. P. & Vierstra, R. D. (1983). Phytochrome: molecular properties and biogenesis. *Philosophical Transactions of the Royal Society of London*.

Rice, H. V. & Briggs, W. R. (1973*a*). Immunochemistry of Phytochrome. *Plant Physiology*, **51**, 939–45.

Rice, H. V. & Briggs, W. R. (1973*b*). Partial characterisation of oat and rye phytochrome. *Plant Physiology*, **51**, 927–38.

Roux, S. J., McEntire, K., Slocum, R. D., Cedel, T. E. & Hale, C. C. (1982). Phytochrome induces photoreversible calcium fluxes in a purified mitochondrial fraction from oats. *Proceedings of the National Academy of Sciences, U.S.A.*, **78**, 283–7.

Rubinstein, B., Drury, K. S. & Park, R. B. (1969). Evidence for bound phytochrome in oat seedlings. *Plant Physiology*, **44**, 105–9.

Saunders, M. J., Cordonnier, M-M., Palevitz, B. A. & Pratt, L. H. (1984). Immunofluorescence visualisation of phytochrome in *Pisum sativum L.* epicotyls using monoclonal antibodies. *Planta*, **159**, 545–53.

Shimazaki, Y., Cordonnier, M-M. & Pratt, L. H. (1983). Phytochrome quantitation in crude extracts of Avena by enzyme-linked immunosorbent assay with monoclonal antibodies. *Planta*, **159**, 534–44.

Smith, W. O. (1983). Phytochrome as a molecule. In *Encyclopedia of Plant Physiology, vol. 16A "Photomorphogenesis"*, eds. W. Shropshire & H. Mohr, pp. 96–118. Berlin: Springer-Verlag.

Thomas, B., Crook, N. E. & Penn, S. E. (1984*a*). An enzyme-linked immunosorbent assay for phytochrome. *Physiologia Plantarum*, **60**, 409–15.

Thomas, B., Penn, S. E., Butcher, G. W. & Galfrè, G. (1984*b*). Discrimination between the red- and far-red-absorbing forms of phytochrome from *Avena sativa L.* by monoclonal antibodies. *Planta*, **160**, 382–4.

Tokuhisa, J. C., Daniels, S. M. & Quail, P. H. (1984). Two distinct molecular species of phytochrome in light grown *Avena sativa*. *Plant Physiology Supplement*, **75**, p. 73. Abs. No. 407.

Verbelen, J-P., Pratt, L. H., Butler, W. L. & Tokuyasu, K. (1982). Localization of phytochrome in oats by electron microscopy. *Plant Physiology*, **70**, 867–71.

Vierstra, R. D. & Quail, P. H. (1982*a*). Native phytochrome: inhibition of proteolysis yields a homogeneous monomer of 124 kilodaltons from Avena. *Proceedings of the National Academy of Sciences, U.S.A.*, **79**, 5272–6.

Vierstra, R. D. & Quail, P. H. (1982*b*). Proteolysis alters the spectral properties of 125 kdalton phytochrome from Avena. *Planta*, **156**, 158–65.

Vierstra, R. D., Tokuhisa, J. G. & Quail, P. H. (1983). A solid phase antibody approach to the subcellular localization of phytochrome. *Current Topics in Plant Biochemistry and Physiology*, pp. 37–50.

Vierstra, R. D., Cordonnier, M. M., Pratt, L. H. & Quail, P. H. (1984). Native phytochrome: immunoblot analysis of relative molecular mass and in vitro proteolytic degradation for several plant species. *Planta*, **160**, 521–8.

D. J. BARBARA and M. F. CLARK

Immunoassays in plant pathology

Introduction

It has long been known that most plant pathogens possess, as part of their structure, specific antigenic determinants in the form of proteins or other antigenic moieties. Recognition of the diagnostic potential of such determinants for both experimental and applied investigations in plant pathology has resulted in a bewildering array of techniques being developed, many derived originally from medical or animal research, which can collectively be referred to as immunoassays. Although early methods lacked sensitivity for antigen detection those developed over the past decade compare favourably in sensitivity with assays based on nucleic acid hybridization procedures in which all or part of the genome of the pathogen is assayed using radiolabelled, specific nucleic acid probes. These two approaches are complementary as they measure different attributes of the pathogen.

Immunoassays are used in plant pathology for three main purposes:

(i) Identification. The antigenic components of a particular pathogen are characteristic and are likely to comprise both organism-specific and group-specific determinants. Thus, pathogens can be classified taxonomically according to their serological relationships. This is especially true of plant viruses, where determination of the serological cross-reactivity of a newly identified isolate with established ones has been suggested as a primary step in establishing their taxonomic status (Hamilton *et al.*, 1981). With many pathogens, even where a group of isolates is considered on the basis of other criteria to comprise a single taxonomic unit or species, they are commonly sub-divided into serotypes.

(ii) Diagnosis. The principle use of immunoassays is for the detection of pathogenic organisms associated with plant disease, either within, or isolated from, affected plant tissues. Although applicable to most pathogens such assays are most widely used with plant viruses where they are often the most sensitive and quickest tests available. Numerically diagnostic assays far exceed those for taxonomic purposes.

(iii) Quantitation. Recently-developed immunoassays, notably solid-phase enzyme- and radio-immunoassays and immunospecific electron microscopy, have enabled direct quantitative measurement of particle or organism numbers, thereby facilitating investigations on distribution of pathogens in plant tissues, environmental effects on numbers of organisms, disease threshold levels and disease control.

Immunological techniques have also been used for various other purposes in plant pathology: for example the localization of pathogens within tissues (e.g. Brlansky *et al.*, 1982); the identification of specific antigens in electrophoretically separated components (e.g. fibril protein of *Spiroplasma citri*; Townsend & Archer, 1983); quantitation of disease-related antigens such as the pathogenesis-related proteins of tobacco (Antoniw *et al.*, 1985).

This review will be largely concerned with the detection and identification of free antigens rather than with the immunolocalization of antigens.

Immunoassay procedures

Many of the immunoassay techniques currently used in plant pathology are outlined below, together with some observations on their relative merits and applicability. Except for those described in the Protocols, details of the procedures are omitted. Such information, and discussions of the theoretical aspects of immunoassays, are given elsewhere (Milne & Luisoni, 1977; Schaad, 1979; Clark, 1981*b*; Van Regenmortel, 1982). Laboratory protocols for various enzyme-linked immunosorbent assays and associated techniques have been described by Clark & Bar-Joseph (1984).

Immunoassays may be divided into three broad categories: those that rely on the formation of insoluble antigen-antibody aggregates which can be seen either directly (e.g. in tube or microprecipitin tests, ring interface tests, gel diffusion assays) or indirectly by using inert particulate markers (e.g. latex and chloroplast agglutination methods) to which one or both reactants are adsorbed, which aid precipitation and enhance the visibility of the reaction; those in which either individual antigen units (such as virus particles), or small numbers of such units, are reacted with antibody, often leading to their being immobilized onto a solid phase, and either visualized directly (e.g. immunospecific electron microscopy) or via an associated label (e.g. enzyme-linked immunosorbent assay); and miscellaneous other assays (e.g. growth inhibition, infectivity neutralization, deformation assays).

1. Aggregation assays

(i) Precipitation/agglutination methods. In these procedures antigen and antibody are mixed and the formation of a visible precipitate indicates a positive reaction. Such reactions are usually referred to as "agglutination" for large, insoluble antigens (roughly the size of a bacterial cell or greater) and as "precipitation" when smaller, soluble antigens are involved. The usefulness of these assays is limited by the relatively large amounts of reactants needed, although these may be reduced in microprecipitin tests, and the need for the antigen to be reasonably free of contaminating host materials which otherwise may obscure the reaction. Agglutination reactions remain useful for large antigens which can be grown in culture and hence are free of host contaminants. Agglutination tests are still popular for detecting and differentiating phytopathogenic bacteria (Trigalet, Samson & Coleno, 1978; Schaad, 1979), e.g. strains of the bacterium *Erwinia chrysanthemi* have been serotyped in this manner (Yakrus & Schaad, 1979).

(ii) Diffusion assays. These are normally carried out in agar or agarose gels, but other media have also been used, e.g. cellulose acetate membranes. Both single (reactants placed in direct contact) and double (separated reactants) immunodiffusion assays have been employed but the latter are usually preferred. The technique is commonly used for detecting and identifying plant viruses and bacteria. Antigen and antibody are allowed to diffuse toward each other and where they balance a precipitin line forms (Van Regenmortel, 1982). The main advantage of diffusion assays over precipitin methods is the possibility of using very crude antigen preparations. The main disadvantage is that they are only applicable to antigens able to diffuse through the agar or other supporting medium. Many icosahedral plant viruses may be tested directly but filamentous viruses and other large antigens must first be fragmented or otherwise reduced in size, by treating with detergents, reducing agents or highly alkaline materials such as pyrrolidine. Where such treatments are employed it is often necessary to use antisera raised against similarly treated immunogens as these frequently exhibit a different serological specificity to those raised against intact particles (Purcifull & Batchelor, 1977).

(iii) Latex agglutination and similar procedures. Several types of inert particle (e.g. bentonite, polystyrene latex beads) have been used to increase the visibility of antigen aggregation reactions. Latex beads are probably the most widely used. The beads are sensitized by passively coating with antibody (the purity of which may be important) and become agglutinated

when mixed with antigen. Either pure antigen or antigen in crude tissue extracts may be used. Sensitization of the latex may be enhanced by a preliminary coating of staphylococcal protein-A (which binds IgG) allowing the use of low titre antisera (Querforth & Paul, 1979; Torrance, 1980). Latex agglutination assays are moderately sensitive (i.e. intermediate between diffusion and binding assays) but they are among the most rapid tests to perform and involve no special apparatus.

(iv) Electrophoretic methods. Immunoelectrophoresis is a powerful analytical method widely exploited for examining mixtures of antigens or for differentiating between related antigens. The antigens are first separated by electrophoresis, normally in agar or agarose, then detected by diffusion against antiserum subsequently placed in troughs cut parallel to the direction of electrophoresis.

Immunoosmophoresis (counter immunoelectrophoresis) involves the convergent electrophoretic movement of negatively-charged antigen and neutral or positively-charged antibody. Positive reactions are seen as a precipitin line. Only sporadic use has been made of this procedure (John, 1965; Clark & Barclay, 1972) although a recent paper describes the detection of double stranded RNA in Fiji disease virus-induced galls by immunoosmophoresis (Rohozinski, Francki & Chu, 1981).

In rocket immunoelectrophoresis antigens migrate electrophoretically into a medium (such as agarose or cellulose acetate) containing 1–10% antiserum. A conical line of precipitation ("rocket") is formed, the area bounded by which is proportional to the antigen concentration. This technique has been used to study the antigenicity of bromoviruses (Rybicki & Von Wechmar, 1981) and as a diagnostic assay for partially-purified cauliflower mosaic virus (Hagen, Taylor & Meagher, 1982).

A two-dimensional form of this type of assay is crossed immunoelectrophoresis (Laurell, 1966). Antigens are separated by electrophoresis in one dimension and this is followed by "rocket" electrophoresis at right angles into an antibody-containing medium. This method is particularly suitable for examining complex mixtures of antigens such as those found during studies on relationships among spiroplasmas (Archer & Best, 1980; Archer & Townsend, 1981) or fungi (Rataj-Guranowska, Wiatroszak & Hornok, 1984).

2. Solid-phase assays

(i) Serologically specific electron microscopy. The earliest combinations of electron microscopy with serology were, in principle, closer to precipitin assays than trapping assays. Initially, pre-reacted mixtures of

antisera and antigens were examined for particle aggregation. However, it was observed that some particles did not aggregate but instead were coated with antibodies (Ball & Brakke, 1968). This coating or "decoration" (Milne & Luisoni, 1977) is now widely applied to the serological identification of particulate antigens, usually viruses. The virus particles are first bound non-specifically to the EM grid, then incubated with antiserum, washed and negatively stained. Decoration of elongate particles is usually easily recognized, but decoration of small icosahedral particles can be more difficult (Rao, Hatta & Francki, 1982). The "visibility" of the coating may be enhanced by the use of electron dense ferritin molecules (Breese & Hsu, 1971) or gold particles as markers, either attached to specific immunoglobulin (Lin, 1984) or via a protein-A intermediary (gold labelled antibody decoration – GLAD; Pares & Whitecross, 1982).

Decoration with antibody facilitates detection of particles in the electron microscope and is especially useful when few particles are present. Double decoration using a second layer of unlabelled anti-species antibodies to increase the "visibility" of particles at low magnifications has also been described (Kerlan, Mille & Dunez, 1981). Some measure of quantitation is possible using arbitrary scoring methods to estimate the degree of decoration of the particles. This technique has been used to assess relationships among polyviruses (Walkey & Webb, 1984). Antibody decoration may also be used to analyse the structure of particles, for example to show the presence of double-stranded RNA in fijivirus particles (Van der Lubbe, Hatta & Francki, 1979) or to demonstrate phenotypic mixing of viral coat protein subunits (Otsuki & Takebe, 1978).

The number of particles adhering to the grid may often be greatly increased by pre-treating the grid with specific antiserum (Derrick, 1973). This procedure is best termed immunosorbent electron microscopy (ISEM; Roberts & Harrison, 1979) to distinguish it from other forms of serologically specific electron microscopy. The observed increase in number of particles binding to treated grids, as compared with untreated ones, can vary from ten to many thousand-fold (Milne, 1980; Kerlan *et al.*, 1981). However, there have been failures which have been ascribed either to the presence of free coat protein "blocking" attachment sites or to the inhibitory effects of serum proteins (Milne, 1980). With purified preparations of virus as many particles were present on untreated grids as on those treated with homologous antiserum but many fewer were found on grids pre-treated with antiserum to an unrelated virus (Leseman, Bozarth & Koenig, 1980).

ISEM is an extremely sensitive technique (capable of detecting virus at concentrations as low as 1–10 ng cm^{-3}; Clark, 1981*a*) and it has been used for example to detect nepoviruses in individual nematodes

(Roberts & Brown, 1980) and potato leaf-roll virus in individual aphids (Roberts & Harrison, 1979). ISEM may also be used for quantitative comparisons using a standard counting procedure (e.g. Roberts, 1980), for example to study serological relationships among luteoviruses (Roberts, Tamada & Harrison, 1980) and isolates of sugarcane mosaic virus (Shukla & Gough, 1984).

Some care is necessary in interpreting the results of all serological electron microscopy. Roberts & Harrison (1979) showed that potato mop-top virus particles were well decorated after two hours treatment with homologous antiserum but were entirely disrupted after four hours. Pares & Whitecross (1983) warned that ISEM increased the apparent proportion of shorter particles in tobamoviruses and obscured the distinction between unimodal and bimodal length distributions. However Roberts & Harrison (1979) found a clearer definition of modal lengths than had been observed previously with potato mop-top tobamovirus and Adams & Barbara (1982*a*) found no evidence of significant particle breakage or of divergence from normal length with carlaviruses from hop.

(ii) Labelled antibody techniques. A considerable increase in sensitivity for detecting antigens has been obtained using antibodies coupled either with enzymes (most widely used) or with radioactive tracers. The use of such labels to signal the reaction of antibody with antigen has enabled detection of antigens at concentrations as low as 1 ng cm^{-3}. The procedure requires the separation of unreacted, labelled antibodies from the reacting components (heterogeneous assays). Accordingly, solid-phase assays are used almost exclusively as these enable the reactants to be presented sequen-

Fig. 1. Types of ELISA.

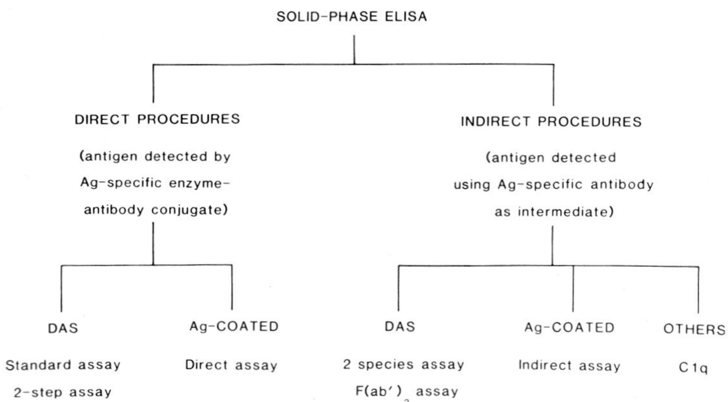

tially at the reaction site, with removal of unreacted components after suitable incubation simply by washing.

Enzyme-linked immunosorbent assay (ELISA) is now probably the most widely used of these serological techniques as it is simple and rapid to perform and is extremely sensitive for antigen detection. Several forms of ELISA are in current use (Fig. 1) and detailed procedures for many of these may be found in Clark & Bar-Joseph (1984), with fuller discussions of the principles and applications of immunosorbent assays in reviews by Bar-Joseph & Garnsey (1981) and Clark (1981*a*, *b*). The procedures may be classified as "direct", in which enzyme-labelled specific antibody is reacted directly with antigen, and "indirect" in which the specific antibody is unlabelled ("native") but is itself reacted with an enzyme-labelled moiety such as an anti-species antibody or protein A.

The basic double antibody sandwich (DAS) procedure, using antibody-coated microtitration plates (Clark & Adams, 1977), has proved remarkably adaptable for a wide range of viruses and other pathogens of different size and morphology; fungi (e.g. *Epichloe typhina*; Johnson *et al.*, 1982); xylem-limited bacteria (e.g. the Pierce's disease bacterium; Nome *et al.*, 1980); spiroplasmas (Clark *et al.*, 1978) and non-cultivated mycoplasma-like (MLO) organisms (e.g. aster yellows; Sinha & Benhamou, 1983); symbiotes (e.g. *Rhizobium*; Kishinevsky & Bar-Joseph, 1978); individual enzymes (e.g. lipoxygenase; Yabuuchi *et al.*, 1982). It has been adopted as the preferred method for plant viruses by research laboratories and plant health monitoring agencies in many countries.

In this procedure (Fig. 2) antigen (e.g. virus) is trapped from the sample by specific antibody previously attached to the solid phase, usually a 96-well microtitration plate of polystyrene or polyvinyl chloride. Trapped antigen is detected by reaction with specific antibody "tagged" with a suitable enzyme label. Presence of the enzyme, and hence of the antigen, is revealed by adding substrate which is converted by the enzyme from a colourless to a coloured compound. Quantitative estimates of the amount of antigen trapped can be made by colorimetrically measuring the rate or extent of substrate conversion.

While the DAS ELISA procedure has proved satisfactory for the majority of indexing applications the need for a separate antibody-enzyme conjugate for each pathogen to be tested can limit certain types of investigation. Variation in antibody titre and in the efficiency of conjugation with enzyme can cause complications with particular combinations of antigen and antibody. Moreover several authors have found that the direct form of DAS ELISA is unsuitable for detecting cross-reactions between some viruses and particular heterologous antisera (Koenig, 1978; Van Regenmortel & Burckhard, 1980; Koenig, Fribourg & Jones, 1979). Such high specificity probably has several different causes, such as lower activity in conjugated as opposed to unconjugated antibodies (Ghabrial & Shepherd, 1980), reduced avidity (Koenig, 1978) or steric hindrance. High specificity has been regarded as beneficial by some authors as it allows differentiation of related antigens (e.g. Barbara *et al.*, 1978) but as a disadvantage by others (see Koenig & Paul, 1982).

For broad spectrum reactivity it may be preferable to employ an "indirect" form of ELISA. The main advantage of indirect procedures is that the usable range of concentrations of detecting antibody is not constrained by the effects of an attached enzyme and may be varied much more than in "direct" procedures, thereby enabling selection of the requisite level of cross-reactivity (Barbara & Clark, 1982). Such assays may be used to determine rapidly the relationships between trapped antigen and a range of antisera (Adams & Barbara, 1982*b*).

Many forms of indirect ELISA have been developed. Sera from two species are sometimes used in combination with an anti-species conjugate,

Fig. 2. Principle of the direct double antibody sandwich ELISA.

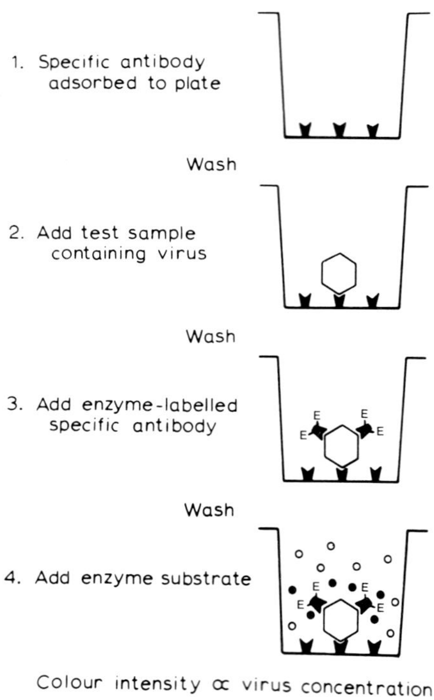

e.g. HADAS ELISA using hens and rabbits – rabbit globulin-sensitized plate/antigen/hen egg antibodies/rabbit anti-chicken conjugate/substrate (Bar-Joseph & Malkinson, 1980). A procedure similar in principle but requiring only a single species antiserum is that based on the use of $F(ab')_2$ fragments of immunoglobulin developed in our laboratory, i.e. $F(ab')_2$ fragment-sensitized plate/antigen/IgG/protein-A-enzyme/substrate (Barbara & Clark, 1982). Simplified indirect ELISA techniques based on the direct attachment of antigen to plates, either actively with glutaraldehyde (Ehlers & Paul, 1984) or passively (Koenig, 1978; Lommel, McCain & Morris, 1982) have been described, i.e. antigen-coated plate/specific antibody/protein-A or anti-species conjugate/substrate. Passive adsorption of virus from crude plant extracts is reported to be reliable for some viruses, e.g. four viruses of carnation (Lommel *et al.*, 1983) but at East Malling some isolates of plum pox virus failed to bind from extracts of *Nicotiana clevelandii* whereas others, indistinguishable by DAS ELISA, did (Barbara & Adams, unpublished results).

Solid-phase radioimmunosorbent assays (RISA), in which the enzyme label is replaced by radioisotope, are directly comparable in principle to ELISA. In such assays radioactivity is counted following release of the bound label with an acidic buffer (Ghabrial & Shepherd, 1980), or counted *in situ* if solid-phase carriers such as polystyrene beads are used (Bryant, Durand & Hill, 1983). However, isotopic labels, although generally about as sensitive as an equivalent ELISA assay, are not widely used in plant pathology, possibly because of the inconvenience or difficulty of handling radioisotopes in many general laboratories.

Labelled antibodies may also be used to identify specific antigens separated from mixtures by various procedures. For example, after electrophoresis in agarose or polyacrylamide gels (PAG) the separated antigens may be transferred to cellulose nitrate membranes and probed with labelled, specific antibodies (see Cuming *et al.*, this volume). Thus, brome mosaic virus antigens were detected following PAG electrophoresis (PAGE) by probing with specific antibodies then a peroxidase-labelled goat anti-rabbit conjugate (with insoluble substrate) (Von Wechmar *et al.*, 1984). A similar technique, employing diazotized paper and isotopically labelled protein-A, has been suggested for the routine detection of plant viruses (O'Donnell, Shukla & Gough, 1982). These assays are likely to have limited potential for routine diagnostic purposes because of low throughput. Nevertheless, they should prove extremely useful in those studies in which the separating power of techniques such as PAGE can be combined with the specificity of immunological detection.

Fluorescent moieties, such as fluorescein and rhodamine, in combination with U.V. light microscopy offer alternative labels which may be used to detect and localize pathogens in tissues (usually in thin sections), e.g. *Verticillium lecanii* in bean rust pustules (Mendgen & Casper, 1980), or to identify serologically and discriminate between morphologically similar structures having different surface antigens, e.g. spores of *Verticillium albo-atrum* (Mohan & Ride, 1983).

3. Miscellaneous other assays

Where biologically active materials are of interest assays which directly affect that activity, such as neutralization of vector-transmitted plant viruses (Gold & Duffus, 1967; Rochow & Ball, 1967), may be used.

Choice of technique

With so many different immunoassay techniques available it is clear that for many purposes any one of several techniques could be used and it is important to select the most appropriate. The final choice will be largely determined by the nature of the problem, the scale of the investigation and the degree of sensitivity and specificity required; a method suited to a large scale field survey may not be so appropriate for detailed comparisons of a few isolates of a pathogen. In selecting the method several important criteria should be considered.

(i) Effects of antigen morphology. ELISA and RISA are generally unaffected by antigen morphology but with ISEM the antigen must be large enough to be visualized and distinctive enough to be recognized. We have found it extremely difficult, for example, to recognize mycoplasma-like organisms (MLO) amongst plant material when trapping the MLO from crude homogenates of infected plant material using antisera with significant

Table 1. *Characteristics of five immunoassay techniques for the detection of plant viruses.*

Technique	Sensitivity ng cm^{-3}	Susceptibility to inhibitors	Quantitative potential	Technical requirements
Immunodiffusion	20,000	low	low	low
Precipitin	2000	high	medium	low
Flocculation	200	medium	low	low
Immune electron microscopy	0.1–1.0	low	high	high
Enzyme immunoassay	1.0–10.0	low	high	medium

anti-plant activity (Clark & Barbara, unpublished results), although others seem to have been successful (Sinha & Benhamou, 1983; Sinha & Chikowski, 1984). Conversely, for gel diffusion assays large antigens must be reduced in size by degradation (e.g. with SDS; Wisler *et al.*, 1982) or by using extracts of the pathogen (Thaveechai & Schaad, 1984). As noted above this may require the production of antisera against the degraded antigen rather than the whole organism.

(ii) Economy of reagents. For large scale surveys a method should be chosen that is economical, especially in the use of antiserum. Precipitin-based assays are generally much less economical of reagents than are solid-phase assays; latex agglutination and similar assays are intermediate.

(iii) Ease of operation. SSEM requires a high degree of technical competence and access to an electron microscope; in contrast double-diffusion and latex agglutination assays require very little equipment. ELISA can also be performed very simply (although precise quantitation and high throughput may require more expensive equipment). Like SSEM, RISA requires expensive apparatus (a radioactivity counter) and it may not be possible to handle radioisotopes easily in all laboratories.

(iv) Sample preparation. Sample preparation may constitute the major time and labour requirement. This input is kept to a minimum when no preparation other than a simple extraction is required. When large numbers of samples are to be examined, as in field-based surveys, even simple extraction procedures can be tedious and time-consuming and assays requiring more elaborate treatments should be approached cautiously. The extreme of simple preparation is the embedding of whole seeds in agar in a single diffusion assay (Von Wechmar *et al.*, 1984) or the use of intact leaf discs in ELISA (Marco & Cohen, 1979). With SSEM the actual examination of samples can be time-consuming and tiring, particularly when performed on a large scale; positive samples may often be recognized in a few seconds but to distinguish confidently between a sample with a low concentration of particles and a negative sample may take several minutes.

(v) Speed. Latex agglutination and ISEM may be used to obtain results quickly whilst most ELISA methods require up to 24 hr (although many samples can be processed simultaneously) and diffusion assays usually require from one to several days.

(vi) Specificity. For many investigations serological specificity is a

consideration of critical importance. Thus, for epidemiological studies conditions givinĝhigh specificity may be appropriate to discriminate among strains of the pathogen. Conversely, serodiagnostic procedures for use in plant quarantine and many field-based pathogen-indexing schemes require broad specificity to reduce the possibility that serological variants are not "missed". Selection of technique and choice of reagents are both factors that can influence specificity.

Classically, serological relationships and antisera specificities have been determined using precipitation-based methods with results being expressed as relative dilution end-points or in terms of serological differentiation indices (Van Regenmortel, 1975). Direct ELISA is considered by some authors to be too specific for determining relationships (e.g. Koenig, 1981). However it has also been described as "especially useful for studying antigenic relationships amongst plant viruses" (Rao *et al.*, 1982). Indirect ELISAs are generally less specific and may be used to study relationships of up to the equivalent of 7 SDI in a precipitin assay (Van Regenmortel, Nicolaieff & Burkhard, 1980). We have developed procedures based on regression analysis for quantifying data from $F(ab')_2$-based ELISA for which we have found the use of high antibody dilutions to be very satisfactory (Clark & Barbara, in preparation).

The discriminatory power of ELISA can be a powerful tool. Rice, Olsen & Page (1984) readily distinguished two strains of *Rhizobium meliloti* in individual root nodules when studying the optimum proportions of strains in preparations used for dressing seed. Mink (1980) was able to identify the cherry rugose mosaic isolates of prunus necrotic ringspot virus from other less virulent strains using a single antiserum.

(vii) Sensitivity. Clearly the assay employed must be sufficiently sensitive to detect the antigen being studied. Table 1 gives a comparison of approximate relative sensitivities for a number of techniques. In general aggregation assays are less sensitive than binding ones, although a given procedure may be especially sensitive for a particular antigen e.g. immuno-osmophoresis for double-stranded RNA (Rohozinski *et al.*, 1981). One advantage of high sensitivity is the ability to process composite samples and hence reduce the number of assays required. Several authors have discussed the rationale of interpreting the results from bulked samples (e.g. Clark, 1981*b*; Moran, Garrett & Fairweather, 1983).

Antisera

The characteristics of an antiserum will affect its usefulness for a particular purpose. One obvious pre-requisite is that it must have been raised against an appropriate antigen (for example see above regarding SDS degradation). Less easily defined are the effects of the injection and bleeding schedule where, for example, early bleeds tend to give more specific antisera but of lower avidity than late bleeds. In general antisera will be required that exhibit low activity against plant proteins but this may be less important in some assays e.g. ISEM. Where it is difficult to produce antisera free from significant anti-plant activity it may be possible to reduce such unwanted activity by extensive cross-absorption of the antiserum; for example antisera having very high anti-plant activities have been used successfully after cross-absorption to detect clover phyllody MLO in an indirect ELISA (Clark, Barbara & Davies, 1983).

Recently, there has been great interest in the production of monoclonal antibodies (McAb) for use in plant pathology, particularly virology and their potential is currently being evaluated in many laboratories throughout the world. Methods for producing McAb are discussed elsewhere in this volume (see Chapter by Galfrè and Butcher, this volume). There are three principal areas in which such antibodies may be of advantage to pathologists. First, they are extremely site-specific and have been used, for example, to detect single amino-acid substitutions in the coat protein of mutants of tobacco mosaic virus (Al Moudallal, Briand & Van Regenmortel, 1982) and to differentiate isolates of this virus previously found to be indistinguishable with antisera (Briand, Al Moudallal & Van Regenmortel, 1982). However Martin & Stace-Smith (1984) working with potato leaf-roll virus found that the three antibody clones they studied had the same specificity as a polyspecific antiserum. The second advantage of using McAb, as found by Martin & Stace-Smith (1984) is their ability to reduce the problem of non-specific (anti-plant) reactions, which are frequently important with potato leaf-roll virus. The third advantage is that a continued and reproducible supply of the antibody can be ensured. It is probable that McAb will also be extremely important in obtaining good serological detection of many pathogens which are difficult to prepare in purified form for the production of antisera.

Recent applications

This chapter has been largely concerned with the mechanics of immunoassay techniques but in general it is not techniques but their applications which are important. Serology has long been used for taxonomic purposes, during which time techniques may have altered but the net result of classification remains largely unaltered and such work will continue to be important. Diagnostic serology using techniques such as gel diffusion has been applied on quite large scales for selected purposes but a more general application has been restricted because of low sensitivity, unsuitable antigen morphology and expense in terms of reagents. Such limitations have largely disappeared with the development over the last decade of extremely sensitive techniques, primarily ELISA and ISEM, which are also quick, efficient in reagent usage and, in the case of ELISA, well suited to processing large numbers of samples. These attributes have meant that serology is now much more widely used as illustrated in this section.

The adaptability and sensitivity of ELISA has meant that field surveys can now be undertaken on a much wider scale than hitherto. Citrus tristeza (CTV), an important disease of citrus induced by an aphid-borne virus, used to be indexed biologically on a large scale by grafting to tristeza-sensitive Mexican lime seedlings but such tests were expensive, time-consuming and permitted only limited throughput. SDS-immunodiffusion was tried but was applicable only to the new growth (Garnsey, Gonsalves & Purcifull, 1978). ELISA was investigated and proved to be suitable (Bar-Joseph *et al.*, 1979). It has been applied successfully in the field to the detection of tristeza in fruit pedicels and in the "button" (calyx and receptacle) of fruit (Bar-Joseph, Sacks & Garnsey, 1978). When samples of five fruits were collected from commercial 350 kg containers and tested for CTV by ELISA there was good correlation between the degree of infection in the orchard as estimated from these samples and the actual level found by testing leaf samples. When only low levels of infection are expected and sensitivity of the assay is sufficient much larger samples may be used. Moran *et al.*. (1983) could detect one potato virus X-infected leaf amongst 1,000 healthy leaves and one potato virus S infection amongst 300. They also showed mathematically that ten samples, each containing 100 leaves were sufficient to reliably determine infection rates of less than 1% in a crop. Thus they were able to test 29,000 plants from 29 plantings in only 290 assays. Bulk sampling of this kind is extremely useful in testing seed-lots for seed-borne viruses. Falk & Purcifull (1983), for example, were able to test in three weeks as many lettuce seed stocks for the presence of lettuce mosaic virus as took 4.5 months by the more usual infectivity test to indicator hosts. The interpretation of such results requires some care as although serological positives may often correlate well with the proportion of infected seedlings found (Bowers & Goodman, 1979; Falk & Purcifull, 1983; Bharathan *et al.*, 1984) this is not always so (Lister, Carroll & Zaske, 1981). An extreme example is that of apple mosaic in horse chestnut; viral antigen was present in a large proportion of immature embryos in seed from infected trees but was lost on maturation and no seed-transmission resulted (Sweet & Barbara, 1979).

High sensitivity is also an advantage where only small amounts of tissue are available as in the detection of pathogens in individual vectors. Such studies are likely to gain in importance in epidemiological programmes. ELISA has been used to detect virus in individual aphids, e.g. the non-persistently transmitted cucumber mosaic virus (Gera, Loebenstein & Raccah, 1978) and beet western yellows which is persistent (Hewings & D'Arcy, 1984). However, success in such investigations is unpredictable and in the case of cucumber mosaic quoted above very much higher levels of virus were detected than was expected. Even for persistently-transmitted viruses it is often necessary to bulk a number of insects together for testing, e.g. beet curly top virus (Mumford, 1982) and potato leaf-roll virus (Clarke, Converse & Kojima, 1980). Potato leaf-roll has been detected in individual aphids by ISEM (Roberts & Harrison, 1979) and when dealing with insects this assay often seems slightly more sensitive than ELISA. It may be easier to detect pathogens in larger insects. Bean pod mottle virus could be detected by ELISA in its beetle vector at a dilution equivalent to one infected insect amongst 250 healthy individuals (Ghabrial & Schultz, 1983). Indeed, virus concentrations were much higher in the insects than in the plants they had been feeding on. The presence of this virus in its vector was a reliable predictive guide to the extent of disease in the soybean crop. By contrast, although rice ragged stunt virus could be detected in individual planthoppers twice as many insects were serologically positive as actually transmitted the virus (Hibino & Kimura, 1982). As with seed transmission, considerable care is always necessary in attempting to correlate serologically positive results with biological activity.

Tests other than ELISA and ISEM can also prove useful. In addition to rice ragged stunt virus several other rice viruses were present in individual planthoppers in sufficient quantity to be detected by the relatively insensitive latex flocculation test (Omura *et al*, 1984).

Conclusions

We have illustrated the wide range of immunoassays available to the plant pathologist and indicated factors to be considered in selecting the most appropriate technique. Each has a role to play but there is no doubt that the development and adoption of highly sensitive assays such as ELISA and ISEM have transformed serology over the last decade and have revolutionized the plant pathologist's, especially the virologist's, approach to the immunological identification of pathogens and the diagnosis of plant disease.

PROTOCOLS

I. Indirect ELISA for viruses using $F(ab')_2$ fragments of immunoglobulin
This method permits the utilisation of a single antiserum and discriminates between trapping and detecting antibodies with enzyme-labelled protein A. This reacts specifically with the Fc portion of whole immunoglobulin used as second (detecting) antibody whilst antigen is trapped by antibodies from which the Fc portion is previously removed by pepsin digestion (Barbara & Clark, 1982). The technique is suitable for use with many pathogens but the recipes below assume the use of a plant virus occurring in reasonable concentration and IgG from a homologous antiserum raised in a rabbit. The assay is a solid-phase assay performed in polystyrene or polyvinyl chloride microtitration plates. Plate quality varies both between manufacturers and categories of usage. Choose a category of plate designed specifically for enzyme immunoassay to obtain reproducible results. For each combination of antigen and antiserum it is necessary first to establish the optimum concentrations of reagents. Refer to Fig. 3 for the layout of the test plate.

The first 4 steps involve the preparation of $F(ab')_2$ fragments; steps 5–16 involve the choosing of suitable assay parameters.

1. Dialyse IgG (1 – 5 mg cm^{-3}) against 0.07 M sodium acetate buffer, pH 4.0, containing 0.05 M sodium chloride (usually 4–6 h is sufficient).

Fig. 3. Suggested lay-out of a test plate for $F(ab')_2$ ELISA.

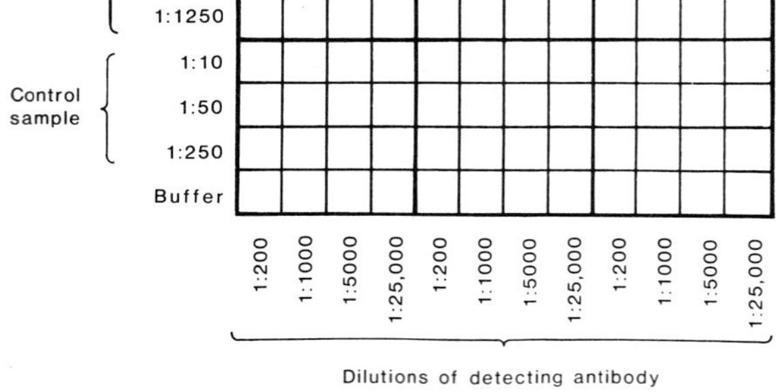

2. Digest IgG overnight with pepsin at 37 °C (e.g. 45 $\mu g\ mg^{-1}$ IgG of 1:10,000 grade from the Sigma Chemical Company, Poole, Dorset, U.K.)[a].
3. Dialyse against three changes of phosphate-buffered saline (PBS; 8.0 g sodium chloride, 1.44 g disodium hydrogen phosphate dihydrate, 0.2 g potassium dihydrogen phosphate, 0.2 g potassium chloride all in 1 litre distilled water, pH 7.4[b].
4. Store at 4 °C with 0.02% sodium azide as preservative[c].
5. Coat the surface of the wells by charging with 200 μl $F(ab')_2$ fragments in 0.05 M sodium carbonate buffer, pH 9.6, at appropriate concentrations (Fig. 3). Seal with clingfilm and incubate at 30 °C 2–4 h.
6. Empty plate by inverting and wash by flooding with PBST (PBS with 0.05% Tween-20). Stand for 3 min.
7. Empty and repeat twice.
8. Charge wells with 200 μl appropriate dilutions of infected and healthy sample extracted into PBST with 2% polyvinylpyrrolidone (M.W. 44,000) plus 0.2% ovalbumin[d]. Seal and incubate overnight at 4 °C.
9. Empty and wash as above.
10. Charge wells with 200 μl whole IgG in PBST (see Fig. 3 for dilutions). Seal and incubate at 30 °C for 3 h.
11. Empty and wash as above.
12. Charge wells with protein-A/horseradish peroxidase conjugate suitably diluted. Seal and incubate as in step 10[e].
13. Empty and wash as above.
14. Add 200 μl substrate[f].
15. After 30 min at room temperature stop the reaction with 50 μl of 3 M sulphuric acid and agitate to mix.
16. Record results visually or by reading at 450 nm if a suitable spectrophotometer or plate reader is available[g].

II. Immunosorbent electron microscopy: ISEM, for plant viruses

Stable antigens with distinctive morphology may be trapped serologically and visualized directly in the electron microscope. It is again assumed that a plant virus and homologous antiserum raised in rabbit are available. Rod-shaped viruses are generally more easily visualized than small icosahedral ones and the former are recommended for the beginner. Many minor variations of the procedure have been reported but the following is suitable for investigating new virus-antiserum combinations.

1. Coat grids with antibodies by floating them on drops of the antiserum diluted in 0.05 M sodium phosphate buffer, pH 6.5, at room temperature for 30 min[h].
2. Wash grids by floating on three changes of distilled water, 2 min per wash.
3. Extract sample by grinding in sodium phosphate buffer (ca. 1:10; w/v).
4. Float coated grids on drops of extract as above for 30–60 min.
5. Wash as above.

6. Negatively stain with 5 to 10 drops and dry[i].
7. Examine in electron microscope for virus particles at a magnification of 18,000 – 35,000, depending on the morphology of the virus[j].

Notes

a If other preparations are used it may be necessary to adjust the amount of pepsin. (Incomplete digestion will show up as strong reactions in wells with no antigen added in the actual assay.)

b It may be more convenient to prepare a stock solution at ten times concentrated.

c The assay is very economical of reagents so that only a small amount need be kept this way. Excess should be stored frozen (avoid repeated freezing and thawing) or preferably freeze dried.

d In many instances these two additives will not be necessary as they are intended to reduce non-specific reactions with "difficult" plant material such as fruit tree leaves.

e See manufacturers recommendation (usually 1:1000) in PBST.

f Stock solution of 10 mg cm^{-3} 3,3′,5,5′-tetramethylbenzidine in dimethylsulphoxide; dilute 1:100 in 0.1 M sodium acetate adjusted to pH 6.0 with citric acid and add hydrogen peroxide to a final concentration of 0.006% just prior to use.

g Suitable dilutions of reagents which give the best discrimination with low non-specific reactions can now be chosen. For routine assays the procedure above is followed except that only single appropriate dilutions of reagents and samples are used.

h Dilutions of 1:500 or 1000 are suitable for most antisera. The drops (ca. 50 μl) should be dispensed onto parafilm stretched over a glass slide to prevent the drops spreading and the incubation done in a moist atmosphere. Avoid antisera which have been stored with added glycerol.

i The most suitable stain will depend on the virus but 2% uranyl acetate is likely to be useful.

j Control grids coated with pre-immune serum (or one to an unrelated virus) should also be prepared.

KEY REFERENCES

Clark, M. F. & Adams, A. N. (1977). Characteristics of the microplate method of enzyme-linked immunosorbent assay for the detection of plant viruses. *Journal of General Virology*, **34**, 475–83.

Derrick, K. S. (1973). Quantitative assay for plant viruses using serologically specific electron microscopy. *Virology*, **56**, 652–3.

REFERENCES

Adams, A. N. & Barbara, D. J. (1982*a*). Host range, purification and some properties of two carlaviruses from hop (Humulus lupulus): hop latent and American hop latent. *Annals of Applied Biology*, **101**, 483–94.

Adams, A. N. & Barbara, D. J. (1982*b*). The use of F(ab′)2-based ELISA to detect serological relationships among carlaviruses. *Annals of Applied Biology*, **101**, 495–500.

Al Moudallal, Z., Briand, J. P. & Van Regenmortel, M. H. V. (1982). Monoclonal antibodies as probes of the antigenic structure of tobacco mosaic virus. *EMBO Journal*, **1**, 1005–10.

Antoniw, J. F., White, R. F., Barbara, D. J., Jones, P. & Longley, A. (1985). The detection of PR(b)-protein and TMV by ELISA in systemic and localized infections of tobacco. *Plant Molecular Biology*, **4**, 55–60.

Archer, D. B. & Best, J. (1980). Serological relatedness of spiroplasmas estimated by enzyme-lined immunosorbent assay and crossed immunoelectrophoresis. *Journal of General Microbiology*, **119**, 413–22.

Archer, D. B. & Townsend, R. (1981). Immunoelectrophoretic separation of spiroplasma antigens. *Journal of General Microbiology*, **123**, 61–8.

Ball, E. M. & Brakke, M. K. (1968). Leaf-dip serology for electron microscopic identification of plant viruses. *Virology*, **36**, 152–5.

Barbara, D. J. & Clark, M. F. (1982). A simple indirect ELISA using $F(ab')_2$ fragments of immunoglobulin. *Journal of General Virology*, **58**, 315–22.

Barbara, D. J., Clark, M. F., Thresh, J. M. & Casper, R. (1978). Rapid detection and serotyping of prunus necrotic ringspot virus in perennial crops by enzyme-linked immunosorbent assay. *Annals of Applied Biology*, **90**, 395–9.

Bar-Joseph, M. & Garnsey, S. M. (1981). Enzyme-linked immunosorbent assay (ELISA): principles and applications for the diagnosis of plant viruses. In *Plant diseases and vectors: ecology and epidemiology*, ed. K. Maramorosch & K.F. Harris, pp. 35–9. New York and London: Academic Press.

Bar-Joseph, M., Garnsey, S. M., Gonsalves, D., Moscovitz, M., Purcifull, D. E., Clark, M. F. & Loebenstein, G. (1979). The use of enzyme-linked immunosorbent assay for detection of citrus tristeza virus. *Phytopathology*, **69**, 190–4.

Bar-Joseph, M. & Malkinson, M. (1980). Hen egg yolk as a source of antiviral antibodies in the enzyme-linked immunosorbent assay (ELISA): a comparison of two plant viruses. *Journal of Virological Methods*, **1**, 179–83.

Bar-Joseph, M., Sacks, J. M. & Garnsey, S. M. (1978). Detection and estimation of citrus tristeza virus infection rates based on ELISA assays of packing house fruit samples. *Phytoparasitica*, **6**, 145–9.

Bharathan, N., Reddy, D. V. R., Rajeshwari, R., Murthy, V. K., Rao, V. R. & Lister, R. M. (1984). Screening peanut germ plasm lines by enzyme-linked immunosorbent assay for seed transmission of peanut mottle virus. *Plant Disease*, **68**, 757–8.

Bowers, G. R. Jr. & Goodman, R. M. (1979). Soybean mosaic virus: infection of soybean seed parts and seed transmission. *Phytopathology*, **69**, 569–72.

Breese, S. S. Jr. & Hsu, K. C. (1971). Techniques of ferritin tagged antibodies. In *Methods in Virology*, vol. **V**, ed. K. Maramorosch & H. Koprowski, pp. 399–422. New York and London: Academic Press.

Briand, J. P., Al Moudallal, Z. & Van Regenmortel, M. H. V. (1982). Serological differentiation of tobamoviruses by means of monoclonal antibodies. *Journal of Virological Methods*, **5**, 293–300.

Brlansky, R. H., Lee, R. F., Timmer, L. W., Purcifull, D. E. & Raju, B. C. (1982). Immunofluorescent detection of xylem-limited bacteria in situ. *Phytopathology*, **72**, 1444–8.

Bryant, G. R., Durand, D. P. & Hill, J. H. (1983). Development of a solid-phase radioimmunoassay for detection of soybean mosaic virus. *Phytopathology*, **73**, 623–9.

Clark, M. F. (1981*a*). Immunoassays in plant virology. In *Immunoassays for the 80s*, ed. A. Voller, A. Bartlett & D. Bidwell, pp. 481–94. Lancaster: MTP Press.

Clark, M. F. (1981*b*). Immunosorbent assays in plant pathology. *Annual Review of Phytopathology*, **19**, 83–106.

Clark, M. F. & Adams, A. N. (1977). Characteristics of the microplate method of enzyme-linked immunosorbent assay for the detection of plant viruses. *Journal of General Virology*, **34**, 475–83.

Clark, M. F., Barbara, D. J. & Davies, D. L. (1983). Production and characteristics of antisera to *Spiroplasma citri* and clover phyllody-associated antigens derived from plants. *Annals of Applied Biology*, **103**, 251–9.

Clark, M. F. & Barclay, P. C. (1972). The use of immuno-osmophoresis in screening a large population of Trifolium repens L. for resistance to white clover mosaic virus. *New Zealand Journal of Agricultural Research*, **15**, 371–5.

Clark, M. F. & Bar-Joseph, M. (1984). Enzyme immunosorbent assays in plant virology. In *Methods in Virology*, vol. **VII**, ed. K. Maramorosch & H. Koprowski, pp. 51–85. New York and London: Academic Press.

Clark, M. F., Flegg, C. L., Bar-Joseph, M. & Rottem, S. (1978). The detection of Spiroplasma citri by enzyme-linked immunosorbent assay (ELISA). *Phytopathologische Zeitschrift*, **92**, 332–7.

Clarke, R. G., Converse, R. H. & Kojima, M. (1980). Enzyme-linked immunosorbent assay to detect potato leafroll virus in potato tubers and viruliferous aphids. *Plant Disease*, **64**, 43–5.

Derrick, K. S. (1973). Quantitative assay for plant viruses using serologically specific electron microscopy. *Virology*, **56**, 652–3.

Ehlers, U. & Paul, H. L. (1984). Binding of viruses from crude plant extracts to glutaraldehyde-treated plates for indirect ELISA. *Journal of Virological Methods*, **8**, 217–24.

Falk, B. W. & Purcifull, D. E. (1983). Development and application of an enzyme-linked immunosorbent assay (ELISA) test to index lettuce seeds for lettuce mosaic virus in Florida. *Plant Disease*, **67**, 413–6.

Garnsey, S. M., Gonsalves, D. & Purcifull, D. E. (1978). Rapid diagnosis of citrus tristeza virus infections by sodium dodecyl sulphate-immunodiffusion procedures. *Phytopathology*, **68**, 88–95.

Gera, A., Loebenstein, G. & Raccah, B. (1978). Detection of cucumber mosaic virus in viruliferous aphids by enzyme-linked immunosorbent assay. *Virology*, **86**, 542–5.

Ghabrial, S. A. & Schultz, F. J. (1983). Serological detection of bean pod mottle virus in bean leaf beetles. *Phytopathology*, **73**, 480–3.

Ghabrial, S. A. & Shepherd, R. J. (1980). A sensitive radioimmunosorbent assay for the detection of plant viruses. *Journal of General Virology*, **48**, 311–7.

Gold, A. H. & Duffus, J. E. (1967). Infectivity neutralization – A serological method as applied to persistent viruses of beets. *Virology*, **31**, 308–13.

Hagen, T. J., Taylor, D. B. & Meagher, R. B. (1982). Rocket immunoelectrophoresis assay for cauliflower mosaic virus. *Phytopathology*, **72**, 239–42.

Hamilton, R. I., Edwardson, J. R., Francki, R. I. B., Hsu, H. T., Hull, R., Koenig, R. & Milne, R. G. (1981). Guidelines for the identification and characterization of plant viruses. *Journal of General Virology*, **54**, 223–41.

Hewings, A. D. & D'Arcy, C. J. (1984). Maximizing the detection capability of a beet western yellows virus ELISA system. *Journal of Virological Methods*, **9**, 131–42.

Hibino, H. & Kimura, I. (1982). Detection of rice ragged stunt virus in insect vectors by enzyme-linked immunosorbent assay. *Phytopathology*, **72**, 656–9.

John, V. T. (1965). A microimmuno-osmophoretic technique for assay of tobacco mosaic virus. *Virology*, **27**, 121–3.

Johnson, M. C., Pirone, T. P., Siegel, M. R. & Varney, D. R. (1982). Detection of Epichloe typhina in tall fescue by means of enzyme-linked immunosorbent assay. *Phytopathology*, **72**, 647–50.

Kerlan, C., Mille, B. & Dunez, J. (1981). Immunosorbent electron microscopy for detecting apple chlorotic leaf spot and plum pox viruses. *Phytopathology*, **71**, 400–4.

Kishinevsky, B. & Bar-Joseph, M. (1978). Rhizobium strain identification in Arachis hypogaea nodules by enzyme-linked immunosorbent assay (ELISA). *Canadian Journal of Microbiology*, **24**, 1537–43.

Koenig, R. (1978). ELISA in the study of homologous and heterologous reactions of plant viruses. *Journal of General Virology*, **40**, 309–18.

Koenig, R. (1981). Indirect ELISA methods for the broad specificity detection of plant viruses. *Journal of General Virology*, **55**, 53–62.

Koenig, R., Fribourg, C. E. & Jones R. A. C. (1979). Symptomatological, serological and electrophoretic diversity of isolates of Andean potato latent virus from different regions of the Andes. *Phytopathology*, **69**, 748–52.

Koenig, R. & Paul, H. L. (1982). Variants of ELISA in plant virus diagnosis. *Journal of Virological Methods*, **5**, 113–25.

Laurell, C. B. (1966). Quantitative estimation of proteins by electrophoresis in agarose gels containing antibodies. *Analytical Biochemistry*, **15**, 45–52.

Lesemann, D. E., Bozarth, R. F. & Koenig, R. (1980). The trapping of tymovirus particles on electron microscope grids by adsorption and serological binding. *Journal of General Virology*, **48**, 257–64.

Lin, N. S. (1984). Gold-IgG complexes improve the detection and identification of viruses in leaf dip preparations. *Journal of Virological Methods*, **8**, 181–90.

Lister, R. M., Carroll, T. W. & Zaske, S. K. (1981). Sensitive serologic detection of barleystripe mosaic virus in barley seed. *Plant Disease*, **65**, 809–14.

Lommel, S. A., McCain, A. H., Mayhew, D. E. & Morris, T. J. (1983). Survey of commercial carnation cultivars for four viruses in California by indirect enzyme-linked immunosorbent assay. *Plant Disease*, **67**, 53–6.

Lommel, S. A., McCain, A. H. & Morris, T. J. (1982). Evaluation of indirect enzyme-linked immunosorbent assay for the detection of plant viruses. *Phytopathology*, **72**, 1018–22.

Marco, S. & Cohen, S. (1979). Rapid detection and titer evaluation of viruses in pepper by enzyme-linked immunosorbent assay. *Phytopathology*, **69**, 1259–62.

Martin, R. R. & Stace-Smith, R. (1984). Production and characterization of monoclonal antibodies specific to potato leaf roll virus. *Canadian Journal of Plant Pathology*, **6**, 206–10.

Mendgen, K. & Casper, R. (1980). Detection of *Verticillium lecanii* in pustules of bean rust (*Uromyces phaseoli*) by immunofluorescence. *Phytopathologische Zeitschrift*, **99**, 362–4.

Milne, R. G. (1980). Some observations and experiments on immunosorbent electron microscopy of plant viruses. *Acta Horticulturae*, **110**, 129–35.

Milne, R. G. & Luisioni, E. (1977). Rapid immune electron microscopy of virus preparations. In *Methods in Virology*, vol. **VI**, ed. K. Maramorosch & H. Koprowski, pp. 265–81. New York: Academic press.

Mink, G. I. (1980). Identification of rugose mosaic-diseased cherry trees by enzyme-linked immunosorbent assay. *Plant Disease*, **64**, 691–4.

Mohan, S. B. & Ride, J. P. (1983). Fluorescent-antibody staining of conidia of isolates of Verticillium albo-atrum in relation to their virulence for hops. *Journal of General Microbiology*, **129**, 3581–9.

Moran, J. R., Garrett, R. G. & Fairweather, J. V. (1983). Strategy for detecting low levels of potato viruses X and S in crops and its application to the Victorian certified seed potato scheme. *Plant Disease*, **67**, 1325–7.

Mumford, D. L. (1982). Using enzyme-linked immunosorbent assay to identify beet leafhopper populations carrying beet curly top virus. *Plant Disease*, **66**, 940–1.

Nome, S. F., Raju, B. C., Goheen, A. C., Nyland, G. & Docampo, D. (1980). Enzyme-linked immunosorbent assay for Pierce's disease bacteria in plant tissues. *Phytopathology*, **70**, 746–9.

O'Donnell, I. J., Shukla, D. D. & Gough, K. H. (1982). Electro-blot radioimmunoassay of virus-infected plant sap – a powerful new tool for detecting plant viruses. *Journal of Virological Methods*, **4**, 19–26.

Omura, T., Hibino, H., Usugi, T., Inoue, H., Morinaka, T., Tsurumachi, S., Ong, C. A., Putta, M., Tsuchizaki, T. & Saito, Y. (1984). Detection of rice viruses in plants and individual insect vectors by latex flocculation test. *Plant Disease*, **68**, 374–8.

Otsuki, Y. & Takebe, I. (1978). Production of mixedly coated particles in tobacco mesophyll protoplasts doubly infected by strains of TMV. *Virology*, **84**, 162–71.

Pares, R. D. & Whitecross, M. I. (1983). A critical examination of the utilisation of serum coated grids to increase particle numbers for length determination of rod-shaped plant viruses. *Journal of Virological Methods*, **7**, 241–50.

Purcifull, D. E. & Batchelor, D. L. (1977). Immunodiffusion tests with sodium dodecyl sulfate (SDS)-treated plant viruses and plant viral inclusions. Bulletin 788 (technical), *Agriculture Experiment Stations Institute of Food and Agricultural Sciences*, University of Florida, Gainesville, Florida, USA, 39pp.

Querfurth, G. & Paul, H. L. (1979). Protein A-coated latex-linked antisera (PALLAS): New reagents for a sensitive test permitting the use of antisera unsuitable for the latex test. *Phytopathologische Zeitschrift*, **94**, 282–5.

Rao, A. L. N., Hatta, T. & Francki, R. I. B. (1982). Comparative studies on tomato aspermy and cucumber mosaic viruses. VII. Serological relationships reinvestigated. *Virology*, **116**, 318–26.

Rataj-Guranowska, M., Wiatroszak, I. & Hornok, L. (1984). Serological comparison of two races of *Fusarium oxysporum* f. sp. *lupini*. *Phytopathologische Zeitschrift*, **110**, 221–5.

Rice, W. A., Olsen, P. E. & Page, W. J. (1984). ELISA evaluation of the competitive abilities of two *Rhizobium meliloti* strains. *Canadian Journal of Microbiology*, **30**, 1187–90.

Roberts, I. M. (1980). A method for providing comparative counts of small particles in electron microscopy. *Journal of Microscopy*, **118**, 241–5.

Roberts, I. M. & Brown, D. J. F. (1980). Detection of six nepoviruses in their nematode vectors by immunosorbent electron microscopy. *Annals of Applied Biology*, **96**, 187–92.

Roberts, I. M. & Harrison, B. D. (1979). Detection of potato leafroll and potato mop-top viruses by immunosorbent electron microscopy. *Annals of Applied Biology*, **93**, 289–97.

Roberts, I. M., Tamada, T. & Harrison, B. D. (1980). Relationship of potato leafroll to luteoviruses: evidence from electron microscope serological tests. *Journal of General Virology*, **47**, 209–13.

Rochow, W. F. & Ball, E. M. (1967). Serological blocking of aphid transmission of barley yellow dwarf virus. *Virology*, **33**, 359–62.

Rohozinski, J., Francki, R. I. B. & Chu, P. W. G. (1981). Detection and identification of Fiji disease virus in infected sugarcane by immunodiffusion, immuno-osmophoretic and enzyme-linked immunosorbent assays. *Journal of Virological Methods*, **3**, 177–86.

Rybicki, E. P. & Von Wechmar, M. B. (1981). The serology of the bromoviruses. 1. Serological interrelationships of the bromoviruses. *Virology*, **109**, 391–402.

Schaad, N. W. (1979). Serological identification of plant pathogenic bacteria. *Annual Review of Phytopathology*, **17**, 123–47.

Shukla, D. D. & Gough, K. H. (1984) Serological relationships among four Australian strains of sugarcane mosaic virus as determined by immune electron microscopy. *Plant Disease*, **68**, 204–6.

Sinha, R. C. & Benhamou, N. (1983). Detection of mycoplasmalike organism antigens from aster yellows-diseased plants by two serological procedures. *Phytopathology*, **73**, 1199–202.

Sinha, R. C. & Chikowski, L. N. (1984). Purification and serological detection of mycoplasmalike organisms from plants affected by peach eastern X-disease. *Canadian Journal of Plant Pathology*, **6**, 200–5.

Sweet, J. B. & Barbara, D. J. (1979). A yellow mosaic disease of horse chestnut (Aesculus spp.) caused by apple mosaic virus. *Annals of Applied Biology*, **92**, 335–41.

Thaveechai, N. & Schaad, N. W. (1984). Comparison of different immunogen preparations for serological identification of *Xanthomonas campestris* pv. *campestris*. *Phytopathology*, **74**, 1065–70.

Torrance, L. (1980). Use of protein A to improve sensitisation of latex particles with antibodies to plant viruses. *Annals of Applied Biology*, **96**, 45–50.

Townsend, R. & Archer, D. B. (1983). A fibril protein antigen specific to Spiroplasma. *Journal of General Microbiology*, **129**, 199–206.

Trigalet, A., Samson, R. & Coleno, A. (1978). Problems related to the use of serology in phytobacteriology. *Proceedings of the Fourth International Conference on Plant Pathogenic Bacteria*, Angers. pp. 271–88.

Van der Lubbe, J. L. M., Hatta, T. & Francki, R. I. B. (1979). Structure of the antigen from Fiji disease virus particles eliciting antibodies specific to double-stranded polyribonucleotides. *Virology*, **95**, 405–14.

Van Regenmortel, M. H. V. (1975). Antigenic relationships between strains of tobacco mosaic virus. *Virology*, **64**, 415–20.

Van Regenmortel, M. H. V. (1982). *Serology and immunochemistry of plant viruses*. New York: Academic Press.

Van Regenmortel, M. H. V. & Burckhard, J. (1980). Detection of a wide spectrum of tobacco mosaic virus strains by indirect enzyme-linked immunosorbent assays (ELISA). *Virology*, **106**, 327–34.

Van Regenmortel, M. H. V., Nicolaieff, A. & Burckhard, J. (1980). Detection of a wide spectrum of virus strains by indirect ELISA and serological trapping electron microscopy (STREM). *Acta Horticulturae*, **110**, 107–18.

Von Wechmar, M. B., Kaufmann, A., Desmarais, F. & Rybicki, E. P. (1984). Detection of seed-transmitted brome mosaic virus by ELISA, radial immunodiffusion and immunoelectroblotting tests. *Phytopathologische Zeitschrift*, **109**, 341–52.

Walkey, D. G. A. & Webb, M. J. W. (1984). The use of a simple electron microscope serology procedure to observe relationships of seven potyviruses. *Phytopathologische Zeitschrift*, **110**, 319–27.

Wisler, G. C., Zettler, F. W. & Purcifull, D. E. (1982). A serodiagnostic technique for detecting cymbidium mosaic and odontoglossum ringspot viruses. *Phytopathology*, **72**, 835–7.

Yabuuchi, S., Lister, R. M., Axelrod, B., Wilcox, J. R. & Nielsen, N. C. (1982). Enzyme-linked immunosorbent assay for the determination of lipoxygenase isoenzymes in soybean. *Crop Science*, **22**, 333–7.

Yakrus, M. & Schaad, N. W. (1979). Serological relationships among strains of *Erwinia chrysanthemi*. *Phytopathology*, **69**, 517–22.

GLOSSARY

Adjuvant. A mixture of substances that delays the release of antigen after injection into an animal. The most commonly used is Freund's. Incomplete Freund's contains oil and an emulsifier. The complete adjuvant also contains mycobacteria.

Allotypes. Different forms of a substance defined immunologically and controlled by allelic genes within the species.

Antibody. Antigen specific immunoglobulin.

Antigen. A substance that induces an immune response when administered to an animal.

Antigenic determinant. An area on the surface of an antigen which combines with the antigen-binding site of an antibody.

Antiserum. Serum containing antibodies raised against a specific antigen.

B cell. An immunocompetent lymphocyte which is the precursor of an antibody-secreting cell.

Bacteroid. The specialised form of *Rhizobium* which fixes nitrogen within the cortical cells of legume root nodules.

Capping. Modification of the 5′ terminus of a mRNA.

Complement. A system of factors in serum which is activated by antibody–antigen interactions and which mediates a number of processes, e.g. cell lysis.

Cytoskeleton. Intracellular matrix of structural proteins. The principle known components include tubulin, actin, intermediate filament-type proteins and nuclear lamins and their associated proteins.

Downstream. An element of DNA (or RNA) sequence which is said to be 3′ to any point may be said to be 'downstream' from that point.

ELISA. Enzyme-linked immunosorbent assay. A solid-phase immunoassay employing an enzyme linked to an antigen (if antibody is immobilised) or to an antibody (if the antigen is immobilised) which is used to produce a specific colour reaction.

Endoproteases. Proteolytic enzymes which cleave polypeptides at internal sites.

Enhancers. Elements of DNA sequence associated with some eukaryotic genes. The presence of such elements increases the level of transcription of their associated gene.

Epifluorescence microscopy. Microscopy using excitatory light directed at the specimen from above rather than from below (as in transmitted light microscopy). Thus the light is reflected down on to the specimen through the objective by a suitable mirror, which also allows the emitted fluorescent light (from the object) to pass up the objective to the eyepiece.

Epitope. A single antigenic determinant.

Exonucleases. Enzymes which degrade polynucleotides by 'chewing in' from the end of the molecule.

Hapten. A small molecule whch exhibits no intrinsic immunogenicity but will do so when coupled to a larger antigenic molecule (often a protein, e.g. bovine serum albumin) known as 'carrier'.

Hybridoma. In the context used here, an 'immortalised' cell line formed by the fusion of antibody-producing spleen cells and myeloma cells.

Imune serum (=antiserum).

Immunogenicity. The effectiveness of a substance to induce an immune response.

Immunoglobulins. A family of proteins produced by cells of the B lineage which act as antigen binding receptors. They exist as cell surface receptors and may be secreted into the blood.

Lymphocytes. One class of white blood cells originating from bone marrow stem cells. They are generally found in blood, lymph and lymphoid tissue (spleen, lymph nodes).

Microtubules. Hollow, proteinaceous fibres, about 25 nm in diameter, composed of tubulin. Within cells, their spatial distribution varies depending both on the stage of the cell cycle and the cell type.

Monoclonal antibody. An antibody secreted by a single clone of cells.

Myeloma. A tumour of myelogenous tissue (bone marrow).

Phragmoplast. Two opposing sets of microtubules positioned between two newly formed daughter nuclei. It is closely associated with cell plate deposition.

Plasmacytoma. A tumour of myelogenous tissue presenting the characteristics of a plasma cell.

Plasma cell. Terminally differentiated antibody-forming cell (derived from a B lymphocyte).

Pre-prophase band. Cortical band of microtubules which supplants the interphase microtubule arrays and decays as the spindle is established. It occupies a restricted zone of the cell cortex, foreshadowing the sites at which the cell plate fuses with the mother cell wall.

RIA. Radioimmunoassay. An immunoassay which involves the competition of radioactively-labelled and unlabelled antigen for the antibody.

Serotype. A category in a classification based on immunological (serological) criteria.

Signal peptide. An amino-terminal extension to polypeptides which are destined to pass into or through the membrane system of cells. Signal peptides differ from transit peptides in that they direct the nascent polypeptide into the endoplasmic reticulum before release from the ribosome. Polypeptides bearing signal peptides are thus synthesised by membrane-bound ribosomes.

T cells. Those lymphocytes normally dependent upon the thymus for their development and not directly engaged in immunoglobulin production.

Transit peptide. An amino-terminal extension to polypeptides which are synthesised in the cytoplasm and subsequently transported into an organelle. The peptide is removed during the act of transport. These organellar peptides are synthesised by free ribosomes.

INDEX

3